Happy Wineing!

Beeby L. Kelley
Kathy Woodhouse

Wineing Your Way Across KENTUCKY

Recipes, History, and Scenery

WRITTEN BY BECKY KELLEY

PHOTOGRAPHY BY KATHY WOODHOUSE

Acclaim Press
—Your Next Great Book—

P.O. Box 238
Morley, MO 63767
(573) 472-9800
www.acclaimpress.com

Cover Photography: Kathy Woodhouse
Book & Cover Design: M. Frene Melton

Library of Congress Control Number: 2015901650
ISBN: 978-1-938905-95-7 / 1-938905-95-4

Second Printing 2015
Printed in the United States of America
10 9 8 7 6 5 4 3 2

Contents

Dedications

First, I would like to thank my teachers for helping me, sometimes forcefully, learn all the intricacies and nuances of the English language. Next, I thank Tyler Madison at the Department of Agriculture, Tom Cottrell, the enologist at the University of Kentucky, and all of the wineries and their owners, vintners, employees, and volunteers who shared their stories, their wines, and always taught me something new at each meeting. I need to thank our husbands, Mike Woodhouse, the appointment setter and designated driver, and Tony Kelley, the official taster and storyteller. It has been quite a journey. Last, thanks to my partner in this book, my friend since we were 8 and 9 (but I'm not a year older, just 364 days), Kathy Woodhouse. Our vision in stories and pictures has come to fruition in a beautiful way for all to see. I can't thank all of you enough.

Salut!

Becky Kelley

Not many people get the chance to complete a labor of love with a friend they have had since childhood. I want to thank my friend, Becky Kelley, for sharing this incredible journey as we Wined Our Way Across Kentucky. Thanks, also, to Becky's husband, Tony Kelley, for his (very) honest feedback on our recipes and his sense of humor that kept us laughing from winery to winery. I learned so much from the hard working, passionate Kentuckians, some working two or three jobs so they can live their dreams of being part of this ancient art of winemaking. Each and every winery and vineyard we visited taught us new lessons as we heard the great stories each had to tell. I want to thank all of them for being so open and accepting of us. Finally, I want to thank my husband, Mike, who has stood by me, encouraged me, inspired me, but most of all believed in me. Without his support, this book wouldn't have been possible. I hope you enjoy *Wine*ing *Your Way Across Kentucky* as much as we did.

Cheers!

Kathy Woodhouse

Foreword

This compilation of winery joy is not only a tribute to the growth of the skills of Kentucky winemakers, but also to the perseverance and gusto of the two women who gathered the information. Becky Kelley, the wordsmith, and Kathy Woodhouse, the bubbly photographer, have put together so many vivid details about the wineries and the winemakers across the widely varied terrain of Kentucky that I was delighted. They have captured the essence of the many family efforts that have grown into real wine producing businesses. They also have identified the wonderful characteristics of a rich spectrum of wine varieties and styles.

The descriptions of their visits to all, and I mean every one, of the Kentucky wineries will give you a good feel for the range and variety of the wines available in the Commonwealth of Kentucky, and where to go for the wine or the experiences you might enjoy. The pictures by Kathy complement the observations by Becky in a warm comfortable way, catching images that expand one's understanding and feel for each winery.

It is difficult for the printed page to develop a sense not only of place, but also of wine essences, but this book is almost as good as going there yourself. I hope it motivates you to visit the wineries that catch your eye.

Dr. Tom Cottrell
Enologist, University of Kentucky

Introduction

Hello fellow wine lovers. This book was created to introduce you to all the wonderful wineries of Kentucky. We would like to introduce readers to every wine as well, but since one winery can have a large and changing variety, that's not possible, so we included a small sampling of each. We hope to take the mystery out of wine for those who may not know the supposed "correct" way to enjoy it. The Kentucky Gals who created this book, friends since they were 8 and 9, want everyone to see what a great representation of Kentucky's hardworking people and award winning vintages these wineries contain. We also heard that wine is a great anti-ager, so we hit the road! Some of the wineries have full menus of Kentucky delicacies; a few offer streams for canoeing or fishing, or trails for hiking; several offer overnight accommodations; and many want to assist in creating unique experiences for special events. Whether or not they host events, and how many they can accommodate, is noted on each winery's page. Most importantly, all of them want to facilitate good times with wonderful people and an appreciation of Kentucky crafted wines.

We've called the paths to follow "tours." When people visit a place for the first time, they often take a tour to familiarize themselves with the area and all it holds. Our goal is also to familiarize visitors with each winery and its individuality. Even though each produces a varying degree of the same product, they are all separate and unique. I'm sorry, no questions are allowed on the tour, but visitors can like our Facebook page if they'd like to contact us: www.facebook.com/kywinegals or search for *Wine*ing *Your Way Across Kentucky*, while on Facebook.

Each tour is not a race to finish, but a path to meander. The beautiful scenery, stories, wine, and sense of camaraderie that exists, is part of the full experience every visitor will want to savor. While following the tours, bring some friends, favorite foods, an up-to-date GPS, and a state map. At times, the traveler must follow the directions on a winery's website or map, as a GPS will send you on a wild goose chase, which is sometimes part of the fun.

About the Recipes

The writer and photographer have not taken cooking classes, but have the expertise learned at our Grandmama's knees. However, we truly understand the roles of two income households and have made many recipes simple enough for weeknight meals, and several can be easily converted to crockpot recipes. Before a recipe was added to the book, it was cooked and photographed by our own hands. We had cooking nights with the hubbies and friends. A recipe was not added unless everyone deemed it "book worthy." Believe me, several were left out, even after a second or third experimentation in the cooking lab. The writer, Becky Kelley, and photographer, Kathy Woodhouse, sincerely hope the people exploring with the help of this guide, have a renewed appreciation for the state as they wine their way across Kentucky. We learned, even after having lived here our entire lives, that there is more to this great Commonwealth than we fathomed. Vive! (That's French for cheers!)

Alphabetical List of Kentucky Wineries

- **Acres of Land Winery** · *Richmond, Madison County*
- **Atwood Hill Winery** · *Morning View, Kenton County*
- **Baker-Bird Winery** · *Augusta, Bracken County*
- **Barker's Blackberry Hill Winery** · *Crittenden, Grant County*
- **Black Barn Winery** · *Lexington, Fayette County*
- **Boucherie Winery** · *Scottsville, Allen County*
- **Bravard Vineyards & Winery** · *Hopkinsville, Christian County*
- **Brianza Gardens & Winery** · *Crittenden, Grant County*
- **Broad Run Vineyards & Winery** · *Louisville, Jefferson County*
- **Brooks Hill Winery** · *Brooks, Bullitt County*
- **Camp Springs Vineyard** · *Camp Springs, Campbell County*
- **Castle Hill Farm** · *Versailles, Woodford County*
- **Cave Hill Vineyard** · *Eubank, Pulaski County*
- **Cave Valley Winery** · *Park City, Barren County*
- **CCC Trail Vineyard and Winery** · *Morehead, Rowan County*
- **Cedar Creek Vineyards** · *Somerset, Pulaski County*
- **Chateau Du Vieux Corbeau Winery** · *Danville, Boyle County*
- **Chrisman Mill Vineyard and Winery** · *Nicholasville, Jessamine County*

- **Chrisman Mill Tasting Room** · *Lexington, Fayette County*
- **Chuckleberry Farm Winery** · *Bloomfield, Nelson County*
- **Crocker Farm Winery** · *Franklin, Simpson County*
- **Echo Valley Winery** · *Flemingsburg, Fleming County*
- **Eddy Grove Vineyards** · *Princeton, Caldwell County*
- **Elk Creek Vineyards** · *Owenton, Owen County*
- **Equus Run Vineyards and Winery** · *Midway, Woodford County*
- **First Vineyard** · *Nicholasville, Jessamine County*
- **Forest Edge Winery** · *Shepherdsville, Bullitt County*
- **Generation Hill Winery** · *Alexandria, Campbell County*
- **Glisson Vineyard and Winery** · *Paducah, McCracken County*
- **Grimes Mill Winery** · *Lexington, Fayette County*
- **Hamon Haven Winery** · *Winchester, Clark County*
- **Harkness Edwards Vineyards** · *Lexington, Fayette County*
- **Highland Winery** · *Seco, Letcher County*
- **Horseshoe Bend Vineyards** · *Willisburg, Washington County*
- **Jean Farris Winery & Bistro** · *Lexington, Fayette County*
- **Lake Cumberland Winery** · *Monticello, Wayne County*
- **Lovers Leap Vineyards & Winery** · *Lawrenceburg, Anderson County*
- **Lullaby Ridge** · *Waynesburg, Lincoln County*
- **Mattingly Farms Winery** · *Austin, Barren County*
- **McIntyre's Winery & Berries** · *Bardstown, Nelson County*
- **MillaNova Winery** · *Mt. Washington, Bullitt County*
- **Misty Meadow Winery** · *Owensboro, Daviess County*
- **Noah's Ark Winery** · *Upton, Hart County*
- **Old 502 Winery** · *Louisville, Jefferson County*
- **Prodigy Vineyards & Winery** · *Frankfort, Franklin County*
- **Purple Toad Winery** · *Paducah, McCracken County*
- **Reid's Livery Winery** · *Alvaton, Warren County*
- **Redman's Farm Winery** · *Morning View, Kenton County*
- **Rising Sons Home Farm Winery** · *Lawrenceburg, Anderson County*
- **River Valley Winery** · *Carrollton, Carroll County*
- **Rock Springs Winery and Vineyard** · *Grayson, Carter County*
- **Rose Hill Farm Winery** · *Butler, Pendleton County*
- **Ruby Moon Vineyard & Winery** · *Henderson, Henderson County*
- **Seven Wells Vineyard & Winery** · *California, Campbell County*
- **Sinking Valley Vineyard & Winery** · *Somerset, Pulaski County*
- **Smith-Berry Vineyard & Winery** · *New Castle, Henry County*
- **Springhill Winery** · *Bloomfield, Nelson County*
- **StoneBrook Winery** · *Camp Springs, Campbell County*
- **Talon Winery & Vineyards** · *Lexington, Fayette County*
- **Talon Winery Tasting Room** · *Shelbyville, Shelby County*
- **The Little Kentucky River Winery** · *Bedford, Trimble County*
- **Up the Creek Winery** · *Burkesville, Cumberland County*
- **Verona Vineyards** · *Verona, Boone County*
- **White Buck Vineyards and Winery** · *Morganfield, Union County*
- **White Moon Winery** · *Lebanon, Marion County*
- **Wight-Meyer Vineyard & Winery** · *Shepherdsville, Bullitt County*
- **Wildside Winery** · *Versailles, Woodford County*

Wine is one of those things some people tend to avoid. There seem to be a lot of Wine rules which change from red to white and everything in between. As long as you like the taste, there is no such thing as a "wrong" wine. The soil grapes are grown in, fertilizers used, and amount of rain and sun, along with a myriad of other factors, will effect the taste of the grape and thus the wine. According to many of the winery owners, "A good wine is one people are willing to buy."

Many people like the oak taste that accompanies some of the wines, some do not. Some people prefer sweeter wines, others like the drier ones, and still others prefer neither too dry nor too sweet. Peoples' taste buds are developed differently based on what a person inherits and what food experiences train his or her likes and dislikes. Previous experiences with wine will also guide a person's preference. A person does not have to know all the supposed rules to enjoy wine, just follow your own tastes and raise a glass or two in friendship and good health.

Kentucky's 120 counties hold many wineries, with many more applying for licenses every year. Winding through the hills and hollows of this beautiful state will bring people closer to Mother Nature in her many forms, and introduce them to some hard-working Kentuckians whose passion is to make a product for you to enjoy. Pack a cooler and picnic basket with favorite cheeses, meats, and breads, and let the wineries choose wines according to your taste and food choices. While at a winery with good food and great friends you're not only tasting wines, you're making memories.

The Idea of Dry, Wet, and Moist

For those who may be unfamiliar with the term, Wikipedia describes a dry county as "a county in the United States whose government forbids the sale of alcoholic beverages." This definition is too simplistic for the vast laws written surrounding this issue. Kentuckians of legal drinking age, despite which county in which they live, are allowed to make up to 100 gallons of wine per person 21 and over, with 200 gallons maximum per household. There are some dry counties which, through a vote, allow the sale of alcoholic beverages in upscale restaurants, golf clubhouses, or wineries, but nowhere else within the county. We call these counties "moist." This short explanation cannot begin to encompass all the laws governing alcohol sales in Kentucky. In the summer of 2012, Governor Steve Beshear appointed a task force to try and streamline Kentucky's confusing patchwork of quaint alcohol-based laws and help wineries increase their agricultural products for the public.

Before beginning, a short explanation about the terms, dry, semi-dry, semi-sweet, sweet, berry, and dessert wines, in that order, might be helpful.

- **Dry** · *not sweet. Just as up and down are opposites, dry and sweet are opposites.*
- **Semi-Dry** · *usually a lightly sweet, sparkling wine.*
- **Semi-Sweet** · *denotes a wine that is balanced, not too dry or too sweet.*
- **Sweet** · *this may sound simple, but actually is not. Sweet is often confused with berry wines, though they are not the same. Sugar is converted to alcohol during the wine-making process, so the higher the alcohol content, the drier the wine. The lower alcohol content wines are sweeter because of the natural sugar that remains unfermented.*
- **Berry** · *usually denotes a wine made with fruit other than grapes or with a mixture of fruit and grape wine.*
- **Dessert** · *specialty wines which are fortified in different ways with an alcohol content of 14% or greater. These are made to drink alone or with a dessert.*

Previous experiences with wine will dictate which type a person prefers. Most people trying wine for the first time prefer sweeter wines. After drinking these for a while, a few months or years, depending on how often one tries them, a person's palate will become accustomed to them and be prepared for the semi-sweet and semi-dry wines. Last, a person will come to enjoy the dry wines. If one starts with the dry, he or she may not like it and not explore further. After becoming accustomed to all the types of wine, when going for a tasting, start with the dry and work down to the sweet.

With this information and personal tastes, anyone is ready to wind his or her way around the beauty of Kentucky.

The Wine Tours

● *The Blue Licks Tour*
Cincinnati Area, Northern Kentucky
NEW CASTLE • BEDFORD • CARROLLTON • OWENTON • CRITTENDEN • VERONA • UNION • MORNING VIEW • ALEXANDRIA • BELLEVIEW • CAMP SPRINGS • AUGUSTA • BUTLER • CALIFORNIA

○ *The Lewis and Clark Tour*
Louisville Area, Northern Kentucky
LOUISVILLE • BROOKS • MT. WASHINGTON • SHEPHERDSVILLE

◔ *The History Tour*
Bardstown Area, Central Kentucky
LEBANON • BARDSTOWN • BLOOMFIELD • WILLISBURG

● *The Thoroughbred Tour*
Frankfort Area, Central Kentucky
SHELBYVILLE • LAWRENCEBURG • FRANKFORT • MIDWAY • LEXINGTON • VERSAILLES

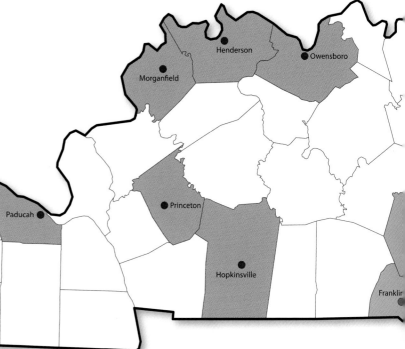

The Stone Fence Tour
Lexington Area, Central Kentucky

LEXINGTON • WINCHESTER • NICHOLASVILLE • RICHMOND • DANVILLE

The Wonderland Tour
Paducah Area, Western Kentucky

OWENSBORO • HENDERSON • MORGANFIELD • PADUCAH • PRINCETON • HOPKINSVILLE

The Cave Country Tour
Bowling Green Area, Southern Kentucky

UPTON • PARK CITY • ALVATON • FRANKLIN

The Lake Cumberland Tour
Somerset Area, South Central Kentucky

EUBANK • SOMERSET • MONTICELLO • BURKESVILLE

The Daniel Boone Tour
Morehead Area, Eastern Kentucky

GRAYSON • MOREHEAD • SECO

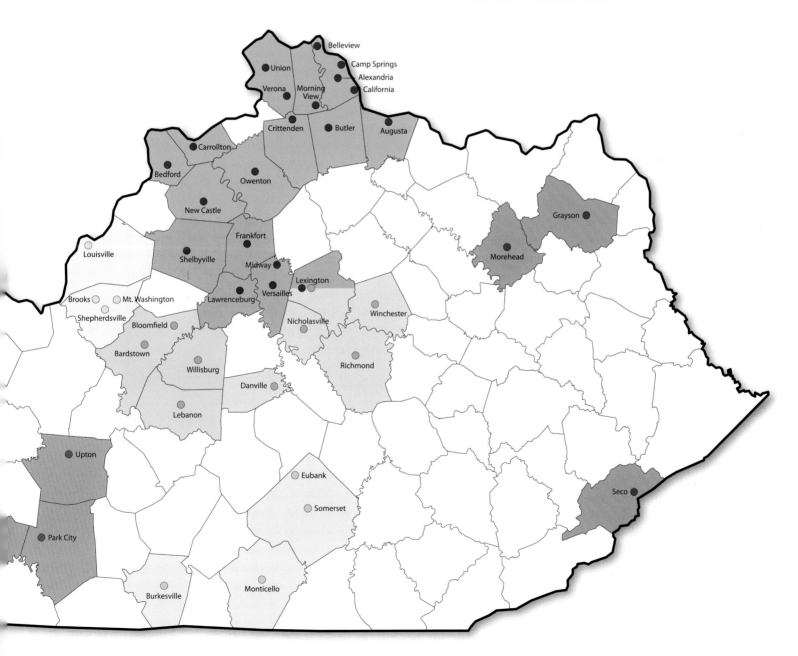

Northern Kentucky Region

CINCINNATI AREA

New Castle • Bedford • Carrollton • Owenton • Crittenden • Verona • Union • Morning View
Alexandria • Belleview • Camp Springs • Augusta • Butler • California

The Blue Licks Tour

*I*n this part of the state, the visitor will find bustling cities and meandering roads, coupled in a pleasing balance of modern and quaint. Many of the drives are close to covered bridges, a once abundant structure that is quickly fading.

In ancient times, salt licks drew mastodons and other animal ancestors to the area. At Big Bone Lick State Park in Boone County, visitors can see their fossils and learn of the history of the area. The park calls itself, "the birthplace of American paleontology," because of the 1807 expedition of William Clark and his brother General George Rogers Clark, who first noted the curiosities there. Native Americans and early pioneers were drawn to this area for the salt, fertile land, the accessibility of fresh water, and abundant game populations.

The Battle of Blue Licks, which occurred on August 19, 1782, is commemorated with a state park in Robertson County. The skirmish that occurred on this site is considered the final battle of the American Revolutionary War. During the Civil War, the Underground Railroad, neither underground nor a railroad, relied heavily on Northern Kentucky, often as the last stop before freedom. Aquariums, art museums, water parks, and shopping are available in Newport and the largest permanent hedge maze in the U.S. can be found in Flemingsburg.

Visitors here will find a large portion of Kentucky history, hospitality, picturesque rolling hills, and, of course, award winning wines.

SMITH-BERRY VINEYARD & WINERY • *New Castle*

THE LITTLE KENTUCKY RIVER WINERY • *Bedford*

RIVER VALLEY WINERY • *Carrollton*

ELK CREEK VINEYARDS • *Owenton*

BARKER'S BLACKBERRY HILL WINERY • *Crittenden*

VERONA VINEYARDS & WINERY • *Verona / Union*

ATWOOD HILL WINERY & VINEYARD • *Morning View*

REDMAN'S FARM & WINERY • *Morning View*

GENERATION HILL WINERY • *Alexandria / Belleview*

CAMP SPRINGS VINEYARD & WINERY • *Camp Springs*

STONEBROOK WINERY • *Camp Springs*

BAKER-BIRD WINERY • *Augusta*

ROSE HILL FARM WINERY • *Butler*

SEVEN WELLS VINEYARD & WINERY • *California*

SMITH-BERRY VINEYARD & WINERY
NEW CASTLE

Gently rolling hills and soft curves greet visitors to this portion of Kentucky. Less mountainous than many parts of the commonwealth, this area off of I-71 is full of the fruits of Kentucky farmers. Drivers will see huge rolls of hay, acres of livestock, and rows of corn, leading to a relaxing place to spend a sunny spring afternoon; Smith-Berry Vineyard and Winery.

In 1981, Chuck Smith purchased 181 acres in Henry County, making them the eighth generation to farm land in this part of Kentucky. The plan was to raise dairy cattle, tobacco, and vegetables, sell eggs and poultry, and just live off the land as traditional farmers for the rest of his life. However, when tobacco began to decline, he started looking for other ways to make the land sustainable. His research showed that the soil and climate of northern Kentucky, very much mirrored some of France's wine regions. He decided to plant grapes, in addition to the cattle, sheep, and organic vegetables he was raising, and start a winery. However, running a winery and farming are only two of the many things happening at Smith-Berry Winery.

"We're a four season winery" explained Chuck Smith. "We don't have any down time here, even in January."

Some of the people who bought chickens and eggs from the long ago are still buying their farm products. "We're a full service business," said Chuck, "but people have been coming so long, we view them as friends, not customers." Mr. Smith hosts and cooks for many events throughout the year and he works hard to make each one individual and special.

"Many weddings are held under the pergola, and then guests move into the event barn for a sit down dinner before a band sets up so guests can dance the night away." In addition to weddings and events, Mr. Smith hosts a once a month cookout and concert series from May to October each year. He sponsors these events in a converted tobacco barn and raises the beef and vegetables served.

Smith-Berry Vineyard & Winery
855 Drennon Road
New Castle, KY 40050
502-845-7091

Open 7 days a week

Events: Yes, indoor seating in the open
event barn, 275 outdoor seating for
up to 1,000.

Smith-Berry has many award winning wines and these accolades came from both local and national competitions. They were recently awarded double gold at the largest contest in the world featuring American wines, the San Francisco Wine Competition. "I couldn't be more proud," Smith said. Until 2012, he'd never submitted his wines to any competitions. That year, he won silver on his Chardonnel and the Zinfandel brought home the bronze from San Francisco.

Whether indoors or out, Mr. Smith can accommodate any party with food, wine, and song. Come and taste a bit of true Kentucky country at Smith-Berry Vineyard and Winery.

Wines

❧ Brother John

This wine, named after family member John M. Berry, is similar to a Syrah. Serve with meaty and mushroom pasta sauces, pizzas, and spicy barbecues.

❧ Norton

This rich wine goes well with spicy Italian, Chinese, or any foods with big flavors.

❧ John Harley

Named after Chuck Smith's father, this wine is a bold tribute. 100% Zinfandel grapes aged in French oak. Raise a glass while watching a glorious Kentucky sunset.

❧ Two Dog White

Semi-sweet and aromatic, this is a great wine to serve with turkey and duck.

Recipe

Sunshine Cheese Ball

8 oz. cream cheese, softened
1 tsp. Dijon mustard
½ tsp. garlic powder
2 Tbls. Smith-Berry Two Dog white wine
½ cup chopped black olives
2 cups shredded Cheddar cheese
2 Tbls. minced parsley or dried flakes
½ cup sunflower seeds

Mix cream cheese, Dijon mustard, garlic powder, white wine, black olives, cheddar cheese, and parsley together. Form into a ball. Roll into sunflower seeds. Keep in the refrigerator until ready to serve.

THE LITTLE KENTUCKY RIVER WINERY
BEDFORD

"A True Country Vineyard and Winery"
– Teresa Weyler –

Usually, tastings at this winery are held outdoors. However, every Saturday in the winter or when the weather is bad, David and Teresa Weyler invite visitors into their space, literally. Their tasting room is their cozy kitchen and the wine is served with a side of Kentucky generosity. The Weylers bought their first piece of ground in Bedford so that David would

have a place to hunt. Together, they kept adding more of the surrounding ground until they had so much, they decided to do something to make it productive.

The land was too hilly to plant many agricultural products, but grapes worked nicely. The hard-working Weylers kept their full-time jobs and jumped into grape growing with both feet. At the same time, they started restoring the original farmhouse on the property. Research showed the house, built in the early 1900s, was first owned by a family named Hunter. A picture album takes visitors through the transition of rebuilding the house and improving the land. People are encouraged to wander the property and commune with nature in the lovely surroundings the Weylers have created. They were restoring the house, learning to make wine, growing grapes, working full-time jobs, and volunteering at other wineries so they could learn all the ins and outs of owning a winery and the art of wine-making.

All of this hard work has resulted in a beautiful winery with secluded wooded spaces for visitors to bring lunch, buy a bottle,

The Little Kentucky River Winery
3289 Highway 421 South
Bedford, KY 40006
502-235-9531

Events: Yes, indoor seating for up
to 100.

and spend an afternoon taking in the quiet allure of the land. In addition, a large pavilion is often filled with people, music, and conversation from spring through fall. All the wood and stone of any structure or building project came from the nearly 650 acres the couple own. Every stone was hand-picked to create the fireplace, porches, and various stone projects on the property.

The wines David and Teresa developed, have special names that reflect Kentucky history, pay homage to the former owners of the property, or honor special people that have come into their lives. Visitors to this winery keep returning to share a bottle and a story or two with the Weylers. You won't want to miss this special attention, interesting stories, and tasty wine. Stop by for a visit soon.

Wines

❧ 1812

A dry Cabernet Franc style wine. The title reflects the War of 1812, fought to enforce treaties with the British Empire and re-enforce America's independence. In this war, 60% of the casualties came from the state of Kentucky and 40% from all other states combined. The name is a reminder to never forget the sacrifice Kentucky made for freedom. It was first bottled in 2012, 200 years after the 32-month war started.

❧ Smoke House Red

A dry chambourcin, is named after an incident that happened to Teresa while renovating the house. Both she and the house accidentally caught fire. Luckily, little damage was done to either Teresa or the house.

❧ Plum Sinful

This is one of the few plum wines at Kentucky wineries. It's very popular, and the visitor will often find this one sold out.

❧ Miss Bessie's Pick

A sweet red raspberry, this wine pays homage to one of the original owners, Miss Bessie Hunter, who was well loved by the whole community.

❧ Dance in the Rain

A semi-sweet white is in remembrance of a special lady who lost her battle with cancer.

Recipe

Rascally Rabbit

2 cups Little River Winery's 1812 wine, divided
3 lb. rabbit, dressed and cut into pieces
1 cup white vinegar
1 dried bay leaf
1 Tbls. dried rosemary
1 tsp. dried thyme
3 cloves
1 tsp. fresh ground pepper
3 pieces bacon
½ cup butter, divided
1 medium sweet onion, chopped
Flour to dredge
Seasoned salt, to taste

Thanks to my friend Teri McLaren at Earthen Vessels Pottery for creating this lovely bowl.

In bowl or large plastic bag. Combine 1 cup of wine with all ingredients except butter, bacon, onion, and flour. Refrigerate for at least 24 hours, but no longer than 48 hours.

When ready to cook, remove rabbit from marinade and allow to drain until no longer dripping.

Fry bacon in large pot that can be put into the oven, such as a Dutch oven. Remove bacon when crispy. Add 2 tablespoons butter and onion, cooking on medium-high until soft and slightly browned. Add remaining butter to the pan. Dredge rabbit in flour mixed with salt and fry in the butter. Once browned, add remaining wine and cook on the top of the stove until it boils slightly. Cover and simmer in a 325° oven for 2 to 3 hours until the rabbit is falling off of the bone. Pour the rabbit and gravy into a bowl and serve with buttered egg noodles, and roasted root vegetables such as carrots, yams, and parsnips.

This is our take on the traditional German dish hasenpfeffer, which literally means pepper hare. If you don't have a hunter in the family, rabbit can be found or ordered at many Kentucky butcher shops and groceries. I found mine at Boone's Butcher Shop in Bardstown, one of my favorite places! Thanks to my good friend Pam LeFaive Myer for introducing it to me.

"Experience a European feel with Kentucky hospitality."
– Vicky and Krasi Georgiev –

Although we arrived on a gray day with snow melting on the ground, the spirit emanating from this winery could not be dampened. A bright welcome flag flapped in the breeze as we drove up the gravel drive to be greeted by 3 Great Pyrenees' Mountain Dogs. Upon entering, the warmth of the room and the kindness of the owners made us feel right at home.

While visiting the Black Sea region on their honeymoon, Vicky and Krasi Georgiev dreamed of opening a winery. Krasi is originally from Bulgaria and the Georgievs decided the gathering places of the traditional Bulgarian wineries would mix well with Kentucky hospitality. They combined the words Bulgarian and hillbillies together and call themselves "Bulhillies."

They planted their first vineyards by hand in 1998, which makes them one of the beginners in the resurgence of the wine industry in the state. The Georgievs call their tasting room "Mehana" which means wine house. The room is simply decorated, in the European tradition, with many items special to the Georgiev Family. A guitar welcomes impromptu concerts from customers, Krasi, or one of their 5 grandchildren. Many of the pictures on the wall were drawn by Krasi and show his artistic side well.

Outside, a grape arbor affords views of the pond, barn, and hilltops all around. Vicky Georgiev, a native Kentuckian, is proud of their agricultural heritage. "100% of our wine is made with Kentucky grown grapes." She's also adamant about food pairings. "Wine is made to enhance the flavors of food and vice-versa." She shows this passion to visitors by setting out favorite cheeses and dips for people to try while sampling wines.

The Georgievs love animals and these are a highlight for many visitors. If you're lucky, you'll get a llama to smile at you.

River Valley Winery
1279 Mound Hill Road
Carrollton, KY 41008
502-750-0594

Events: Yes, indoor seating for up to
40, outdoor seating for up to 200.

The girls, a flock of sheep, were brought in to help with the lawn mowing. The boys, Pyrenees' Mountain Dogs, help to keep the sheep in line. A lone duck also resides on the farm and believes herself to be the matriarch of the other animals. She keeps them in line, cleans the pond of duckweed and moss, and reigns over the farm, while providing eggs for her keep. The wines, animals, hospitality, and views combine to make a trip to River Valley Winery, like a fine wine, an experience to savor.

Wines

❧ Vidal Blanc
A German style wine with grapefruit, pineapple, and citrus flavors. It works well with seafood and poultry, or by itself.

❧ Cayuga White
Crisp and fresh, this wine pairs well with grilled burgers and mild cheeses.

❧ Shiraz
A unique full bodied wine, rich, dark, heavy in tannin with lush texture and ripe fruit flavors. Shiraz is delicious with beef brisket or grilled lamb. Also pairs well with a hearty beef stew or chili.

❧ Loretto
Named after Vicky's mom, this sweet white wine goes well with desserts.

❧ Bobby's Blush
Named after Vicky's dad, this is a great table wine to keep on hand for every day dinners or special occasions.

Recipe

Stacked Tostada

1 ½ cups River Valley Bobby's Blush, divided

1 pkg. corn tortillas

1 large can re-fried beans

1 pkg. fajita seasoning

1 rotisserie chicken (or chicken cooked in slow-cooker for 8 hours), shredded with fork

1 cup favorite salsa

2 medium tomatoes, diced

1 cup sour cream

3 cups shredded lettuce, washed and placed in bowl

3 cups shredded sharp Cheddar cheese

Spray oil in nonstick pan, fry corn tortillas until crispy. In large, deep skillet, pour re-fried beans, ¾ cup wine, and package of fajita seasoning. Heat to boiling, reduce heat to lowest setting. In large skillet, place chicken, salsa, and remaining ¾ cup wine, heat through.

Add tomatoes and sour cream to lettuce and mix together. (This is a handy trick I discovered so that you don't lose all your toppings.)

Place one tortilla on plate, spread 2 tablespoons of beans on top and add 1 tablespoon lettuce mixture. Place another tortilla on top of lettuce mixture; add 2 tablespoons of chicken mixture, top with a tablespoon of shredded cheese.

Eat and Repeat.

ELK CREEK VINEYARDS
OWENTON

"Fine Wines from the great Commonwealth of Kentucky."
– Curtis Sigretto –

While wine*ing* your way to this winery, you may want to follow a map or the directions on the website instead of relying completely on the GPS. We stayed lost for a while on this one, but laughed ourselves silly the whole time with jokes about the GPS-Gabby, Gabriella Page Smith (the name my daughter gave her) plotting to plunge us into a river.

Driving past the expansive vineyard is awe-inspiring. Upon entering Elk Creek Vineyards tasting room, visitors may feel as if they've entered the great hall of a castle or a large hunting lodge. A massive stone fireplace covers one wall, and winter visitors may think they are sitting at a ski lodge. The stone and timber structure is huge with architectural highlights such as a curved stone staircase and tall double doors. This winery is billed as "the largest winery in Kentucky." This is true in the size of the tasting room, the vineyards, and the amount of wine bottled. In addition to the winery, Elk Creek offers lodging, a cafe, a sport and gun club, an outdoor stage, and a beautiful space to have an event or spend a quiet afternoon.

As people sit on the large wrap-around stone patio, the view of the vineyards and a distant pond add to the ambiance of this lovely place. Overnight visitors have several choices from individual rooms to a whole building. The Safari room in the lodge offers a

Elk Creek Vineyards
150 Highway 300
Owenton, KY 40359
502-484-0005

Events: Yes, indoor seating for up to 75, outdoor seating for up to 300.

pool table, bar, entertainment center, and a conference room, plus breakfast is included and massage services are offered.

Visitors to this winery mention sitting on the outside deck and taking the winery tour as their favorite things to do. Many people enjoy the changing array of Elk Creek Café's soups, salads, appetizers, and sandwiches, to pair with their favorite glass of wine. The winemaking and wine cellar tour is not to be missed.

Another unique experience offered at Elk Creek Vineyards is master cooking classes. Eight participants work with a chef to prepare a four course meal and are introduced to wine pairings throughout the class. If you're not interested in learning the cooking methods, but just want to eat, friends and spouses may join the class for the eating portion. Of course, you'll still have to find someone to take the class so you can enjoy the meal, but it shouldn't be hard to find a volunteer.

Whatever your needs, whether it is an event space for 10-300, a place to relax and enjoy some friend time, some music therapy at an indoor or outdoor concert, a guys' or girls' weekend with skeet shooting or fun winery activities, or a quiet couples getaway, Elk Creek Vineyards offers a variety of experiences, all in one place.

Wines

Blackberry Merlot

Sweet and juicy blackberry wine is combined with merlot grapes for a perfect fusion of semi-dry wine. This wine would pair well with salty meats such as prosciutto or country ham, or a savory cheese spread.

Sweet Cranberry

This sweet and tart wine is best served chilled. Sip alone, add to sangria, or serve with a fruit salad at a garden party. It will be a hit!

Niagara

Both citrusy and floral, this sweet wine has a wonderful crisp finish. Great to drink alone after dinner or with a Kentucky summer peach cobbler.

Watermelon Wine

The taste of a Kentucky summer in a bottle. This is the perfect wine to pack in the picnic basket.

Left: Photo credit, April Cole.

Recipe

So good, your family will think you've worked all day.

Country Taters with Style

½ cup Elk Creek Pinot Grigio
Two 14.5-oz. cans sliced potatoes, drained
14.5-oz. can sweet potatoes or yams, drained
1 small carton whipping cream
4 slices American cheese
¾ cup mozzarella cheese
½ tsp. chipotle chili powder

Mix all ingredients except potatoes. Stir to heat and thicken.
Add potatoes, heat through and pour into a bowl.

BARKER'S BLACKBERRY HILL WINERY
CRITTENDEN

"Come taste Kentucky summer in a glass."
– The Barkers –

Winding our way around curves and up and down hills on the way to this winery, we caught glimpses of pure blue streams with waterfalls reflecting the waning winter sun. The beautiful scenery led us to two Kentucky characters, Brenda and Jimmy Barker, who love to laugh, but more importantly, love to make others laugh.

The Barkers welcomed us warmly into their rustic tasting room. "My wife only married me because I have a good garden, and wine." Mr. Barker joked.

"Oh, don't mind him, he never meets a stranger," Mrs. Barker laughed.

In 1990, Kentucky legislation set up a Grape Industry Advisory Committee (GIAC) which helped enact a Kentucky Farm Winery Law. Because of this law, four wineries were established that year, with Jimmy Barker being the first to apply for, and be granted, a permit.

Mr. and Mrs. Barker are also the owners of Barker's U-Pick Blackberry Farm. When asked how he started in the wine making business, he stated. "I was selling a lot of blackberries to people who were making their own wine. I decided I'd try it so none of my crop went to waste."

Although he had plenty of fruit, his first ventures weren't always successful. "I made wine this one time early on, and it wasn't fit to drink. I was worried thinking I'm going to have to pour out 800 gallons of wine. I left it in the tanks and could hear it bubbling. Sometimes, I'd sit and just listen. It was like a concert. After a while, I tried it again, and it was delicious! It became our best-selling blackberry ever."

Barker's Blackberry Hill Winery
16629 Mt. Zion Verona Rd.
Crittenden, KY 41030
859-428-0377

Events: No

Mr. Barker's wine-making philosophy is to get all the fruit as close to home as possible. Of course, all the blackberries come from his 48 and ½ acres. A lady across the street has a grove of cherry trees she doesn't use, so they made an arrangement and barter fruit. He also gets strawberries from a neighbor with a U-Pick Strawberry Farm.

While an ardent lover of Kentucky now, Mr. Barker is a transplant. "I lived in the Virginia Mountains, just across from Kentucky. Ran away when I was 17 and learned construction." As a young man, while helping to build I-75, he found a piece of ground in Kentucky he wanted to acquire. In 1980, he planted his first hillside of thornless blackberries which led to Jimmy Barker's winery path. Visit the Barker's for a laugh and be sure to raise a glass or two to their health while in Northern Kentucky.

Wines

❧ Blackberry

Sweet and refreshing, blackberry wine is a Kentucky favorite. Replace some of the water in your favorite cake recipe with it or mix with powdered sugar and create a glaze or frosting for a sweet treat.

❧ Strawberry

Like biting into early summer, this wine is bursting with flavor. Pour over your favorite pound cake and serve with cut strawberries and whipped cream.

❧ Cherry

Tangy and tart. This wine will make you believe you're eating cherries straight from the tree.

❧ Honey Mead

Hard to find, Mr. Barker knows a bee-keeper to keep this wine flowing in our land of milk and honey. Worth the trip for this wine alone.

Recipe

Take a Break Shortcake

1 lb. of Bing or like cherries
Enough Barker's Cherry Wine to cover cherries
1 ½ cups powdered sugar
Shortcakes or pound cake slices
Vanilla ice cream
Maraschino cherries

Cut cherries in half, remove pits and stems, wash. Cover with wine and soak in refrigerator overnight. Strain cherries, reserving the juice into a saucepan. Bring juice to hard boil for two minutes. Allow to cool slightly, add powdered sugar, stirring to dissolve. Add cherries back to juice and refrigerate for two hours.

Place one shortcake or one slice of pound cake on plate. Top with two tablespoons of cherries with juice. Top with one scoop of vanilla ice cream and/or whipped cream and a maraschino cherry. Take a break and enjoy your shortcake!

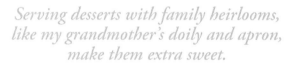

Serving desserts with family heirlooms, like my grandmother's doily and apron, make them extra sweet.

31

VERONA VINEYARDS & WINERY
VERONA / UNION

"From ground to glass, we make a great wine."
– Dan Montgomery –

About 20 years ago, Charlie Payne saw promise in this piece of ground in Boone County and purchased it. He was unsure what it would be used for, but knew he wanted to keep an agricultural heritage and save the historic house on the property.

Coming up the gravel drive, visitors pass the historic George Washington Vest House. Born in Virginia in May of 1757, Vest served in the Revolutionary war under Colonels Thomas Dillard, Andrew Lewis, and Evan Shelby, and General George Rogers Clark. The land was granted to him for his service in the war. The house was built in the 1840s. Current Verona Winery owners, Charlie Payne, Peggy Payne Montgomery and Dan Montgomery, spent many hours restoring the house to make it their home.

As we pulled into the parking lot, the skies threatened rain, but the sun kept pushing through the clouds, creating a halo effect over the tasting barn. Several people, visitors and owners, exchanged hearty greetings with us as we made our way inside, so we were feeling all warm and fuzzy before we ever tasted the wine at Verona.

The owners of Verona Winery decided to use all the natural help they could in the vineyards. "We researched for sheep that would keep the grass down between the vines, but wouldn't be tall enough to eat the grapes and found some from an English bloodline we liked," explained

Verona Vineyards & Winery
13815 Walton-Verona Rd.
Verona, KY 41092
859-485-3544

Events: Yes, indoor seating for up to 75,
outdoor seating for up to 150.

**Verona Vineyards at Rabbit Hash
Satellite Wine Shop and Tasting
Room**
11646 Lower River Rd.
Union, KY 41091
859-322-0487

Events: Yes, indoor seating for up to 20.

Dan Montgomery. "We thought they'd be hard to find, but a breeder lived in Bardstown." Along with the adorable Southdown Babydoll Sheep on the property, the Payne's and Montgomery's rescue horses and dogs.

Dan and his wife Peggy have kept their original careers, so Charlie Payne is in charge of the day to day farm business. The first grapes were planted in the spring of 2004 with the intention of selling grapes to other wineries, but the especially abundant harvest of 2008, steered them in another direction. "We found ourselves with a bunch of grapes and no buyer," Dan said. "So, we decided to make wine." Their first venture into wine-making produced about 400 bottles. For comparison, in 2011 the vineyard produced about 7,000 gallons of wine. With about 5 bottles to a gallon, that's around 35,000 bottles of wine.

The tasting barn at Verona is the place where the family's generosity shines. Friends often gather, but visitors are treated as kin. One member of the family pours the tastings and answers questions with skill, while another may sit at a table talking with visitors as if they're old friends. People may think they've stumbled into a private party or event, but that's just business as usual at Verona Vineyards and Winery in Verona Kentucky. It's filled with friends, spirit, and wine. Enjoy!

Wines

Chardonnay

Available in both oaked and un-oaked versions, this dry white offers citrus and melon flavors that compliment chicken and fish dishes, as well as white pasta sauces.

Cabernet Franc

A soft, easy to drink red wine. Light, peppery, with notes of oak and smoke with a nice leather and fruit finish. Pairs well with Middle Eastern and Greek cuisines.

Zinfandel

Light and fruity with aromas of raspberry and oak, this wine is great to sip alone, add to cheese spreads for a little zip, and pair with a favorite dessert. It's zin-ful.

Riesling

This sweet wine has hints of peach and citrus. Great chilled on a hot Kentucky evening or with a piece of caramel apple pie. Yum!

Recipe

Chicken Franc

2 cups Verona Cabernet Franc
3 Tbls. olive oil, more if needed
2 cloves garlic, minced
6 boneless, skinless chicken breasts
1 Tbls. smoked paprika
1 cup dark brown sugar
1 can chicken broth
16-oz. pkg. egg noodles
4 Tbls. butter
3 Tbls. parmesan cheese
Salt and freshly ground pepper to taste

Heat butter in a large skillet over medium-high heat. Add garlic, stirring often, until tender. Add oil and chicken to the skillet and cook about 10 minutes on each side, or until juices run clear. Push the chicken to one side and sprinkle chicken with paprika. Add brown sugar and wine, stir to dissolve sugar. Cover and simmer for 15 to 20 minutes, basting chicken with sauce every 5 minutes. Season with salt and pepper.

Cook egg noodles according to package directions. Drain and add butter and Parmesan cheese. Serve chicken and sauce over noodles.

ATWOOD HILL WINERY & VINEYARD
MORNING VIEW

"Wines are a journey of taste."
– Julie Clinkenbeard –

We arrived at this winery on a bright sunny afternoon to find cats roaming the grounds and a hearse in the parking lot. We were afraid we'd stumbled upon a private ceremony, until we realized the hearse was reconditioned for sitting patrons and drove customers to several winery tastings. What a way to go! Tasting, that is.

This 6th generation family farm has literally turned over a new leaf, converting their tobacco fields to wine groves when they planted their first grapes in 2005. "We sold the grapes at first," explained owner Julie Clinkenbeard, "but we wanted to stay more hands on with our product." Tobacco farmers plant in small cups inside before transferring the seedlings to the ground once the threat of frost has passed. Throughout the growing season they hoe, top, cut, stake, pick, dry, and then take it to market. With grapes, they felt they didn't get the same feel of the land if they didn't see the production to the end results.

Atwood Hill Winery & Vineyard
1616 Spillman Road
Morning View, KY 41063
859-356-1936

Events: Yes, indoor seating for up to
50, outdoor seating for up to 50.

Visitors to Atwood Hill Winery will often find Julie Clinkenbeard running the tasting room. She enjoys meeting people from all over the state, country, and even world, on her and husband Nelson Clinkenbeard's farm. "Seeing the farm grow, the wine trails developing, and the people willing to travel and try different wines, makes me hope wine will return to its pre-prohibition popularity," Ms. Clinkenbeard stated. At that time, Kentucky was the third largest grape producer in the U.S.

Upon entering the attractive wine tasting house, there is an immediate sense of family. One man in our party, originally from England, laughed heartily at a comment from an Irishman at a nearby table referring to long standing sports rivalries between their two countries. "Oh we like the English individually," he commented in his deep Irish brogue. "It's as a group that we have trouble with them."

Moving outside, visitors take in beautiful vistas from under the enchanting pergola, which supports a substantial grape vine. Sitting there with the sun shining, sipping wine, and hearing several different accents, people can imagine they're on a world tour. Visitors can imagine they hear Italian folk music or a German polka floating in from just over a nearby hillside. Visit Atwood Hill and taste the journey. Salute!

Wines

❧ **Atwood Reserve**

A dry Vidal this crisp white wine is wonderful in or with white pasta dishes, turkey, and ham.

❧ **Kentucky Barrel Red**

A rich ruby color is brought on by the Chambourcin grape. Aged in Bourbon barrels, this wine is perfect for a cookout or with spicy deli-meats.

❧ **Oh Charlie**

A Vidal and Chambourcin blend, this blush is perfect to serve with light summer chicken and tuna salads.

❧ **Paige White**

From the Cuyuga grape, this semi-sweet wine would work well with appetizers.

❧ **Sunset Lynn**

Atwood's version of a white Zinfandel, serve with creamy pasta dishes and grilled fish.

Recipe

Straw-Berry Good Bread

1 cup Atwood Hill Strawberry Wine

3 cups self-rising flour 1 ½ cups sugar 2 tsp. ground
   cinnamon ¾ cup canola oil

3 eggs, beaten 1 tsp. vanilla 2 ½ cups chopped fresh
   strawberries

Two 8-oz. tubs cream cheese spread with
   strawberries

Preheat oven to 325°. Grease and flour two 8-inch loaf pans. Set aside.

Combine flour, sugar, and cinnamon, whisk and make a well in the middle. Add oil, wine, and eggs. Stir together well, scraping the bottom to get all dry ingredients mixed in well. Add vanilla and strawberries and stir slightly to incorporate berries throughout the batter. Divide batter between loaf pans. Spoon cream cheese with strawberries in a line down the middle of the batter and swirl throughout with a knife, it does not need to be even.

Bake 1 hour or until knife inserted in center comes out clean. Cool on rack before removing from pan. Once cool, slice and serve. A sweet white or rosé wine is perfect to serve with this treat.

Redman's Farm, established in 1941, is a 5th generation working family homestead with very little down time, thus the need to call for an appointment. Nestled in a valley of southern Kenton County, drivers follow railroad tracks for a bit before coming to this traditional home farm winery. It's a tourist attraction in itself offering a very popular Halloween themed experience with hayrides, pumpkin patch trips, and a not-too-scary Graveyard tour of a cemetery that sits on the property. They also participate in the Roadside Farmer's Market program, selling their products to the public at the farm or a farmer's market close to their home. In addition, the Licking River borders the property and there's a lake for fishing, so bring a pole and bait. Kids can meet the farm animals or pick fruits, vegetables, and flowers, but the highlight of the visit will be meeting the Redman's and, for the adults, tasting the wine.

Marge and Doug Redman are the head honchos at Redman's Farm, but their daughter Tara Scheidt and her husband Tim run a close second. Mr. Redman learned his winemaking tradition from his grandfather, who used to make wine each year at harvest time as a hobby. Remembering this long ago ritual, Doug Redman began making fruit wines a gallon at a time to see how they would taste. He also played around with different farm produce to see which could be made into wine. He discovered that almost anything will make wine, even pumpkins! "We just experimented with what we thought might taste good," Doug Redman said. "We add wines by what we like." Not being afraid to experiment has been a successful route to winemaking, as they have several people who stop in regularly to buy wine by the gallon.

Redman's Farm & Winery
12449 DeCoursey Pike
Morning View, KY 41063
859-356-2837 *(Call for appointment*

Events: Yes, indoor seating for up to
30, outdoor seating for up to 100.

"It's a toss-up what our best seller would be." Tara explained. "Strawberry and blackberry are both good sellers." That could be because a lot of cooks like to use these varieties of wine as a secret ingredient for cakes and pies. There's just something about adding a little, or a lot, of wine to a recipe that changes the whole way it tastes. It makes cakes more moist and fruit pies tangier.

The Redman's prefer to make all of their wines from home grown fruit or fruit grown close by. They offer strictly berry wines of Blackberry, Strawberry, Apple, and Cherry. They're experimenting with a pumpkin wine. Varieties may change, depending on what other fruits they may grow or obtain from neighbors. While the varieties may change, one thing does not, they're all delicious.

The whimsical decorations in the tasting room, add to the farm feel of the entire experience. An old cook stove stands proudly against one wall, and it's the first thing most visitors see when they enter. A chicken nest with eggs in it, contributes to the ambiance of the farm décor. One really interesting display is a tobacco basket hanging on one wall. It's filled with lighted clusters of grapes in a representation of the transition of moving from the old and embracing the new.

Come and spend an afternoon with the Redman's soon and get wrapped up in their many adventure stories. As they leave the tasting room, visitors might unexpectedly find the sun has gone down.

Wines

❧ Apple
Available in both sweet and semi-sweet varieties, this is a versatile wine to drink by itself or serve with a cheese plate. Heat slightly and add a cinnamon stick for a cozy winter treat.

❧ Blackberry
Considered an ice wine, the berries are frozen to sweeten and intensify the blackberry flavor. Kentucky summer in a glass.

❧ Strawberry
This wine comes in both sweet and semi-sweet choices. This is the wine to sip while sitting on the back porch watching the sun go down. A perfect ending to a Kentucky fall day.

❧ Concord Dry
Reminiscent of eating grapes in Grandpa's garden, this wine is perfect to pair with savory beef dishes.

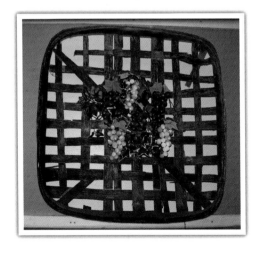

Recipe

H-apple-y Ever After

1 cup Redman's Farm Apple Wine
4 to 5 lb. pork loin, cut in half
2 Granny Smith apples, cored and chopped
2 sweet potatoes, peeled and chopped
½ stick butter
¾ tsp. seasoned salt
½ tsp. pepper
4 tsp. yellow mustard
2 Tbls. flour

In a skillet, melt butter and toss in apples and sweet potatoes. After 3 to 4 minutes, when just beginning to soften, put in slow cooker. In the same skillet, brown pork loin and add to slow cooker. Add wine to skillet and stir to get all the yummy bits, a process also known as deglazing. Add seasoned salt, pepper, mustard, and flour. Stir until smooth and just boiling. Pour over pork, apples, and sweet potatoes. Cook on low for 8 to 10 hours. Remove to dish and spoon more of the gravy over it.

A dish my family likes with this is cabbage. I put the skillet used to brown the pork roast in the fridge, then pull it out when I get home, add chopped cabbage, salt and pepper, 2 tbs. of butter and ¼ cup of the apple wine. Stir and cook about 10 minutes until tender, but still a bit crispy. Serve with the pork, apples, and sweet potatoes.

Some of the best days are when you know dinner is already taken care of. Slow Cooker recipes are so great. It's just a little work in the morning, a little work in the evening, and a delicious dinner for the family to enjoy together.

GENERATION HILL WINERY
ALEXANDRIA / BELLEVUE

"Easy Going, Lovable, and Fun."
– Bill Buda –

Located close to Florence Kentucky with its many activities, Bill Buda's Generation Hill Winery has the notable distinction of being the smallest winery in the state, and that's just fine with him. Before opening in 2010, Buda started wine-making as a hobby, just to have something to do. Now as a retired engineer from Brown and Williamson Tobacco, he is happy to be a part of Kentucky's wine industry. It was his career that actually set him on his wine path.

Buda traveled the world for his job and even lived in England for a time. While traveling, he noticed that everyone, from the smallest cottage to the largest farm, seemed to have a vineyard, yet in America, we didn't share this quaint custom. It charmed him that so many cultures shared the ancient tradition of making their own wines to share with friends and family and the wine bug bit. "Working the vines, learning the correct way to prune and spray to produce a good grape, is highly satisfying." Buda started experimenting by growing different types of grapes to see what would grow best on the 18 acres of vines and fruit trees.

A hot, dry summer is highly prized by grape growers because it yields a lot of product. "This year's weather was perfect," Buda noted. "Look at how beautiful the concords turned out." He let us eat our fill as he proudly showed us around his property. "This was my wife's family's homestead, 4th generation. It used to be a dairy farm and they

Generation Hill Winery
335 Poplar Thicket Road
Alexandria, KY 41001
859-694-1561

Events: No

Generation Hill Tasting Room
315 Fairfield Avenue
Bellevue, KY 41073
859-816-7142

Events: Yes, indoors for up to 12.

grew fruit trees." He's kept the fruit-growing tradition with his grape vines, but visitors to this winery will not be able to buy wine here, it's in a dry county. There is, however, a tasting room in the city of Bellevue, but Buda likes to get out and see many people.

"I sell my wine at farmer's markets and festivals," Buda explained. People can find him at Bellevue Farmer's Market most Wednesdays and Saturdays. He also likes Winchester's Sip & Stroll Festival in late summer and the MainStrasse Festival in Covington the third weekend in May. "I only make 4-5,000 bottles a year." He might run out of some kinds, so only selling at certain venues allows many people to try a lot of different types of his wines.

His wine labels, designed by him or his wife, are something they enjoy doing together. She often finds hilarious mishaps to record for all to see. If you have some time, stop by the Bellevue Tasting Room for a visit. While there, ask Mr. Buda how he got the nickname, "Mr. Grape Jeans."

Wines

☞ Chambourcin

Best seller and gold medal winner at the Indy International Wine Competition, this deep purple wine is fruity with just a hint of spice. Pair with hearty beef dishes.

☞ Kentucky Can-Can

Semi-sweet and estate grown, this red wine is great for creating a sangria or pair with your favorite red pasta sauce.

☞ Mellow Dog

A blend of Seyval Blanc and Viognier, this wine is smooth and crisp with a touch of apple flavor. Serve in or with your favorite chicken dish.

☞ Lorelei of Kentucky

A semi-sweet medium bodied white, this wine is grown from Kentucky Riesling grapes. It adds depth to white pasta sauces or white gravies and pairs well with a cheese and deli tray.

Recipe

Coq au Vin (Chicken with Wine Sauce)

Generation Hill Mellow Dog Wine
4 chicken breasts (preferably bone in)
3 cans cream of mushroom soup

Use an electric skillet, large covered frying pan or Dutch oven. Skin and cut away excess fat from chicken. Heat the skillet and coat with cooking spray. Sear the chicken and pour in the mushroom soup. Fill one empty soup can with Mellow Dog Wine and pour over chicken and soup mixture. Cover and simmer for 40 minutes or until chicken is thoroughly cooked. Serve with rice and your favorite green vegetable such as broccoli.

CAMP SPRINGS VINEYARD & WINERY
CAMP SPRINGS

Trees converge over the road creating a colorful arch for visitors on their way to this winery. It feels as if it is out in the country, but there are many modern tourist attractions within a few miles of this vineyard including the World Golf Academy and several art galleries in Newport, Kentucky, as well as a fantastic children's museum and zoo in Cincinnati. Visitors may need to stay a few days to take in all of those, plus spend time with the father and sons team of Lonnie, Chris, and Kevin (primary chemist) Enzweiler.

In 1963, Lonnie Enzweiler bought the farm on which the winery sits. Chris and Kevin grew up there, doing all the chores that agricultural work requires. Camp Springs is the definition of a family farm winery. In 2005, with a desire to make the land productive once again, the Enzweilers first planted 200 Vidal Blanc vines. "The vines look like little sticks at first," Kevin explained. "It's very satisfying seeing them green up and produce fruit." His dad Lonnie added wryly, "Wine connoisseurs ask questions we don't know how to answer. We just want to make wine people will want to drink." This family is a great representation of Kentucky's hard-working farmers trying to make a living off the land. Pictures around the tasting room show the littlest to the eldest Enzweiler's helping in the field. Everyone contributes in some way to make them a success.

They grow many of their grape varieties, but getting some of the fruit to make other types of wine happens in a variety of ways. "We had a neighbor with an abundant crop of pears. He couldn't give them away so he picked 15 or 16 bushels and brought them to us," Kevin said. "We decided to try making pear wine." Although this might sound like a simple idea if you have all of the equipment, a lot of research, science, and math go into creating good wines. Kevin mentioned the pH balance, acidity,

Camp Srings Vineyard & Winery
6685 Four Mile Road
Camp Springs, KY 41059
859-448-0253

Events: Yes, indoor seating on two
levels for up to 100.

and residual sugars as factors that all influence how a wine becomes a satisfying product for all involved.

Their favorite part of owning a winery, once harvest is over, is they get to relax and enjoy their visitors. Lonnie explained, "We've had people from Japan, Argentina, Venezuela, Spain, Germany, and all over America," Lonnie explained. "You could hear Spanish, Portuguese, English, and German all spoken at the same time." When the family started their winery, they never considered that their land would become a meeting place for the world. Be sure to visit the Enzweiler's soon at their winery or the Ft. Thomas Farmer's Market. You may come for the wine, but you'll return for the hospitality.

Wines

Chambourcin
Dry red wine that compliments grilled steaks and burgers, and red pasta sauces.

Merlot
Semi-Sweet, this is not your typical Merlot. Serve with grilled tuna or swordfish and a tomato salad.

Vidal Blanc
Estate grown, this versatile wine comes in sweet, semi-sweet, and semi-dry varieties. Serve with poultry or mild cheeses such as a smoked mozzarella or mild Cheddar.

Four Mile Blend
A blend of Chambourcin and sweet Vidal Blanc, this wine pairs well with a deli-meat and cheese tray.

Raspberry
Sweet as a Kentucky summer evening, this wine pairs well with peach or chocolate desserts.

Blackberry
Sweet and memorable, this wine is great for sipping slightly chilled, or paired with your favorite cheesecake or lemon pie.

Recipe

Vintage Fudge

½ cup Camp Springs Raspberry wine
4-oz. stick butter
2-oz. package chocolate chips
16-oz. package powdered sugar
Nuts, optional

In a double boiler, melt butter and chocolate chips until smooth. In a bowl, mix powdered sugar with wine until smooth and paste-like. Pour melted chocolate into the bowl with wine/powdered sugar mixture and stir until well blended. Pour into an 8x8 glass baking pan. Refrigerate for 1 hour and cut into 1-inch pieces.

Oh Fudge!
You got your wine in my chocolate!

51

STONEBROOK WINERY
CAMP SPRINGS

"Back to our roots."
– Bonnie and Dennis Walter –

In this area of Northern Kentucky, the hills and valleys resemble wine growing regions of Germany. Because of this, many German immigrants settled here seeking religious and/or political freedoms and economic opportunities. Most of those who came before the Civil War (1861-1865) were farmers seeking fertile land. We have many things for which to thank these early immigrants. They established the first kindergartens, introduced the tradition of the Christmas tree, and gave us foods still popular today such as hot dogs and hamburgers. (wikipedia.org/wiki/German_American)

In the late 1860s, the Walter family first came to this area known as the "Rhine of America." The original homestead, still owned by the Walter's family, has seen many changes through the years, in order to remain sustainable.

According to Stonebrook's website, the first Martin ancestor established the farm in 1870, building a blacksmith shop on the property. The shop produced wagons, carriages, surries, and other forms of early transportation. Two hay wagons that can still be seen on the farm, were built at that same blacksmith shop. Sev-

eral buildings from the original farm remain functional today including the original 1870s' homestead. The tasting room is a remodeled 1890s' farmhouse, where many of the tools and

StoneBrook Winery
6570 Vineyard Lane
Camp Springs, KY 41059
859-635-0111

Events: Yes, indoor seating for up to
   65, outdoor seating for up to 65.

household items from that time are on display. When cars and trucks replaced coaches and freight wagons, the farm had to change with the times. Through the years, the Walter's family has raised pigs, vegetables and fruit, and became dairy farmers. At present, in addition to the winery and commercial vineyard, the family raises beef cattle.

Current owners Bonnie and Dennis Walter enjoy living and working on the farm. "It's a challenge," said Bonnie, but "the accomplishment is very satisfying." She went on to explain, "overcoming weather conditions, spring frost or rains and summer drought, while still producing a successful product is very satisfying." Her favorite part, however, is the people. "I get to meet so many people from all over the world and I run into old friends from childhood, people I went to school with, but we lost touch over the years."

This newest part of the Walter homestead, the winery, was opened in 2005. "We started growing grapes for other wineries, and then decided to try making our own wine. The work wasn't a surprise," said Bonnie. As farmers, they're used to working hard on a daily basis. "But the marketing was." Taking the product to the people and advertising were new to them. They currently have twenty varieties of wine, all developed with care and precision. Visit soon and taste how the Walter's have sustained their roots.

Wines

☞ **Vidal Blanc**

Light and Fruity, this wine tempts the palate with citrus and pear notes. It's ideal with poultry or fish, but compliments many strong cheeses such as sharp cheddar.

☞ **Raspberry**

Smooth and bursting with flavor, this medium bodied sweet wine is the perfect after dinner treat. Drink alone or sip with a plain cheesecake.

☞ **Chambourcin**

This dry red wine is a Beaujolais Nouveau style, meaning bottled only 6-8 weeks after harvesting. It pairs well with roast beef or chili.

☞ **Pomegranate**

This wine has won medals from several national and international competitions. Sweet and tart, it tastes great added to a frozen drink or fruit punch.

Recipe

Peach Balls

1 ¼ cups plus 2 tsp. StoneBrook Peach wine, divided
18-oz. pkg. white cake mix
Two 3-oz. boxes peach gelatin
3 egg whites
1 tsp. vanilla
⅓ cup canola oil
15-oz. container ready-made icing (white or cream cheese)
20-oz. white candy coating or almond bark
A variety of colored sugar crystals
Food coloring

Preheat oven to 350° and grease a 9x13 pan. In a large bowl, combine cake mix, 1 box gelatin, egg whites, 1 ¼ cup wine, vanilla, and oil. Mix on low speed until blended, then beat on high speed for two minutes. Pour into pan and bake for 30-35 minutes. Cool cake completely. In large bowl, crumble cake. Add 4 teaspoons Jell-O, container of icing and two teaspoons of wine. Mix thoroughly, form into two-inch balls, and put them on cookie sheets covered with wax paper. Melt 20 oz. white candy coating or almond bark in double boiler. When melted, spoon over the peach balls. While still wet, decorate with sprinkles. Place small amounts of candy coating in a cup and add a drop of food coloring for different colors.

"Enjoy Tasting History"
– Dinah Bird –

Visitors will catch glimpses of hillside houses, black barns, and many tourist attractions as they wind their way around this area of Kentucky. Signs for Evergreen Hills, the largest hedge maze in the U.S., dot the landscape; and Split Rock Conservation Park, which holds programs in ecology, archaeology, geology, and local history, is nearby. The historic jewel of this region, however, is Baker-Bird Vineyard and Winery.

Baker-Bird gets its unusual name from the German tradition of naming a winery after its builder, Baker, and the current owner, Bird. This area along the Ohio River, close to Cincinnati, was known as the "American Rhineland" because the topography resembled Germany's Rhine region so closely it drew many German immigrants. Abraham Baker Sr., a revolutionary war soldier, was deeded the land, now part of Bracken County, as thanks for his service to this country. His son, Abraham Baker Jr. built the winery in the early 1850s. The massive, 12,000 square feet, stone building literally makes visitors stop and stare in wonder. It was built before the time of electricity and power tools, an amazing feat by some hardworking and ingenuitive forefathers.

The owner of this winery, Dinah Bird, has a mission in life, to preserve this building and the stories it contains. Along with her passion for wine and the wine-making process, she has a gift for telling history. "This is the oldest commercial estate winery

Baker-Bird Winery
4465 Augusta Chatham Road
Augusta, KY 41002
606-756-3739

Events: Yes, indoor seating for up to
100, outdoor seating for up to 150.

in America," she explained as she pointed out several documents pertaining to the building and its first owners, the Bakers. "In the winter, you can still see the original planting areas on the hillsides. Civil War soldiers wrote home about eating the grapes from those vineyards." Visitors to Baker-Bird will be inspired by the building, enthralled by Ms. Bird's tour, her knowledge of the grounds, and the history contained within the thick stone walls of this winery.

When the street just in front of the winery, Route 19, Augusta Chatham Road, was due to be built, the plan was to bulldoze the building. Luckily, a group stepped in and had the property placed on the National Historic Register, saving it. Dinah Bird and her husband of 33 years, Martin Westerfield, bought the winery in 2002. On Good Friday of 2006, they planted their first grape vines, determined by what grows best in this region. "Vidal Blanc is the most popular white wine grape grown in Kentucky," Ms. Bird explained. "The vines do well in this area because of the old saying; 'Grapes like to see water, but not get their feet wet.'" Ms. Bird welcomes all visitors to come to Baker-Bird and "taste the history." It's an experience they will never forget.

Wines

- ### Vidal Blanc
  A wonderfully flavorful sweet white, this wine is best chilled and would complement an array of cheeses or a plate of fruit.

- ### Ruby Hawk Rosé
  A delicious blend of Cayuga White, Vidal Blanc, and Marchal Foch, this one is very popular.

- ### Chambourcin
  Deliciously full and spicy, this slightly oaked wine would accompany red pasta dishes with Italian sausage very well.

- ### Frappe Julep
  A sparkling wine, this is perfect to serve at a special occasion, or just to sip on the back deck as the sun goes down on a hot summer day. Gather friends and share the good time!

Recipe

You're Going to Need a Bigger Boat, Scallops

½ cup Baker-Bird Chardonelle
8-12 sea scallops, frozen
2 Tbls. butter
1 orange peeled and divided in half

Place frozen scallops in skillet, turn on medium high heat. Turn and break apart, cooking until a fork goes all the way through them. Drain off water and add butter. Sear scallops on both sides. Remove to plate and keep warm. To the pan, add wine and squeeze the juice from ½ of the orange into the skillet, stirring until reduced and beginning to thicken, about 3 to 4 minutes. Pour sauce over scallops, sprinkle with remaining orange slices for garnish, and serve.

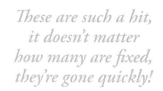

These are such a hit, it doesn't matter how many are fixed, they're gone quickly!

ROSE HILL FARM WINERY
BUTLER

"For the love of the farm."
– Chuck and Jenny Beetz –

One of the best parts about traveling this great Commonwealth to visit wineries is meeting the people who invest so much time and energy to this agricultural passion. Chuck and Jenny Beetz and Jenny's sister Betty Pyles, hosted us on a lovely spring day at their wonderful Rose Hill Farm Winery.

The farm was previously owned by Betty and Jenny's father, but the family ties go back much further in history. They are the fifth generation of the T. J. Campbell family to farm the land since it was obtained in the mid-1800s. "Our great-great-grandfather, Thomas Jefferson (T.J.) Campbell, was born in a little house that used to sit right about there," Betty proudly showed off the spot to us soon after we arrived. "We lived in that white house right up there," Jenny pointed out. For the Beetz's, their line from past to present hasn't always been strictly on the farm, but they always knew that's where they wanted to "retire."

Chuck laughed softly at the word retire because with two separate businesses and all the work of the winery, he's busier now than ever before. "When Dad died," Betty explained, "it was in his will that the farm should be auctioned." There were six children and he thought this would be the fairest way to decide how it should be divided.

"When the bidding at the auction got high," Jenny said, "many of us children were crying, thinking we were going to lose the farm, which was unthinkable." Chuck kept bidding until the farm belonged to the family once again. He and Jenny chose a ridge with expansive views for themselves and the other family members

Rose Hill Farm Winery
199 Highway 17 North
Butler, KY 41006
859-640-8239

Events: Not presently, but there are
future plans to offer events.

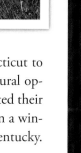

divided the remaining acreage. Afterwards, Chuck retired from his job in Connecticut to move where they always knew they would end up and started looking for agricultural opportunities for the land. They'd always wanted to make wine, so in 2006, they planted their first grapes and tried their hand at winemaking, "as a hobby." They decided to open a winery, determined to grow as much of their fruit as possible and get the rest from Kentucky. They have succeeded in this mission, making Rose Hill a Kentucky Proud winery.

Rose Hill Farm gained its name by accident. Wild roses used to grow all over the hillside, close to the original farmhouse and people started identifying it by this distinctive feature. The name became so popular in the community that when a post card arrived in 1907 addressed to "Rose Hill Farm" they adopted the name and have kept it ever since. They are in the process of being able to sell wine at the farm. Until then, you can pick up their wonderful array of wines at Pendleton Farmer's Market every Saturday May-October, in Louisville at St. Andrew Liquors, Friendly Market and Party Town in Florence, Spirit House in Falmouth, Campbell Shell Liquors in Grants Lick, and Alexandria Carryout in Alexandria.

Wines

🌱 **Rose Red Chambourcin**

A slightly sweeter version of their traditional dry Chambourcin, this wine is perfection in a glass.

🌱 **Seyval Blanc**

A semi-sweet white, this crisp, full-bodied wine has notes of citrus that pair well with pork and poultry.

🌱 **Concord**

This grape originated in Concord, Massachusetts, but adapts to Kentucky soil well. It creates a sweet wine that is wonderful to sip alone or pair with a fruit pie.

🌱 **Apple Wine**

Grown from Kentucky apples, this sweet wine is delicious served slightly chilled or add some cinnamon and nutmeg and heat slightly for a warm winter treat.

Recipe

Don't Bust My Chops

2 ½ cups Rose Hill Farm Winery
   Seyval Blanc, divided
6 thick cut pork chops
Pork flavored, or favorite flavor,
   stuffing mix
1 stick salted butter, divided
1 can French onion soup
2 Tbls. corn starch
Seasoned salt and freshly ground
   pepper, to taste

Mix 2 cups of the wine and ½ stick of the butter in sauce pan. Bring to boil, remove from heat and add stuffing mix. Sprinkle seasoned salt and pepper on pork chops and brown in skillet with 2 tablespoons of the butter, adding more as needed. Make a cut lengthwise through each pork chop, creating a pocket. Fill each chop with stuffing and place in baking pan. Place any remaining stuffing around the chops. In skillet, add remaining ½ cup wine. Mix corn starch into French onion soup, add to skillet, stir to thicken. Pour evenly over chops and bake at 350° for 30 minutes.

SEVEN WELLS VINEYARD & WINERY
CALIFORNIA

"Come visit us among the rolling hills of rural northern Kentucky."
– Doug and Greg Wehrman –

Looking out at the vineyard as you drive up to this lovely cabin-inspired tasting room, visitors know something wonderful waits inside. Seven Wells Vineyard and Winery began as a family project. "My brother Doug saw an article on vineyards in Kentucky," owner Greg Wehrman explained. Their father, Bill Wehrman, bought the original farm in 1973 and raised cattle and hay for over 20 years. By the early 2000s, the cattle had been sold and Bill and his son Greg were looking for a way to make the land profitable. "When my brother saw that," Greg said. "We decided we'd plant 100 vines and if they died, it was a sign we should try something else." Instead, Doug and Greg made a last minute decision to plant 400 vines and most lived with high yields. "I tell people who think about growing grapes or opening a winery, don't plant more than 100 plants the first time. Then you'll get a feel for how much work it is."

The Wehrman's original plan was to sell grapes to wineries, but the amount of work changed Greg's mind. "It was so labor intensive that we figured we might as well try the winery too." It's a good thing they did, as Seven Wells now contains five acres of vineyards and produces twelve high-quality wines.

To prepare for making wine, Greg took Tom Cottrell's wine-making class at the University of Kentucky. "He's a real asset, Kentucky's lucky to have him." Mr. Cottrell's name popped up at many of the wineries throughout the state.

Seven Wells Vineyard & Winery
1223 Siry Road
California, KY 41007
859-816-0003

Events: Yes, indoor seating for up to
30, outdoor seating for up to 50.

The name "Seven Wells" came from a comment Greg's father Bill made while they were trying to come up with a winery name. "There was once seven homesteads on the land, each with its own well." Both brothers and father agreed that it sounded like a fitting name.

Seven Wells is one of the newer wineries in Kentucky, obtaining its winery license in 2009. At first, they decided to only grow grapes other wineries in the region had grown successfully. Vidal Blanc does very well in this part of the country and many wineries offer a Vidal Blanc or Vidal Blanc blend. It's also successful because the taste is satisfying and goes with so many types of food. Chambourcin is also very popular for its ability to grow and its taste. A newer variety they produce, however, is Greg's favorite.

"We tried a Noiret a few years ago." Noiret is a hybrid developed and named at Cornell University. Researchers at the New York State Agricultural Experiment Station officially released it on July 7, 2006 (wikipedia.org/wiki/Noiret). "It's my favorite to sip," Greg explained.

From the beginning, the vineyard and winery has been a family affair with Bill and Greg handling most of the research and work, and Greg's brother Doug volunteering after his full-time job. Sadly, the patriarch Bill Wehrman passed away, but his sons carry on his tradition of hard work and proudly create a homegrown agricultural product. Visit Seven Wells soon and taste the pride. It's more than worth the trip.

Wines

❧ Cabernet Franc

Sophisticated dry with a full-bodied texture, this wine finishes with a spice and slight oak taste. Serve with marinara sauce pastas, pizzas, and lamb dishes.

❧ Traminette

Holding the characteristics of its parent Gewürztraminer, this spicy and full-flavored semi-dry white goes well with baked ham and Asian dishes.

❧ Three Wishes

A sweet wine made from Cayuga grapes; this is the perfect wine for a cocktail party. Pair with bacon appetizers, creamy pasta sauces, and Mediterranean spiced foods.

❧ Raccoon Red

Named because both humans and raccoons seem to prefer the Chambourcin and Dornfelder grapes used to make this wine, this sweet red is a bestseller.

Recipe

Quiche

½ cup Seven Wells Three Wishes white wine
1 ready-made pie crust
8 eggs
¼ cup shredded sharp Cheddar cheese
¼ cup shredded white American cheese
½ tsp. onion powder
¾ cup fresh baby spinach, torn into pieces
½ cup diced ham

Preheat oven to 350°. Place crust in fluted pan or 9-inch round cake pan. To keep a crispier crust, I brown a pie crust slightly before adding fillings. Mix remaining ingredients and pour into crust. Bake for 20 to 30 minutes or until center is set.

I close my eyes and make a wish;
hoping I prepared a beautiful dish.

Northern Kentucky Region

LOUISVILLE AREA

LOUISVILLE • BROOKS • MT. WASHINGTON • SHEPHERDSVILLE

The Lewis and Clark Tour

It's fitting to begin this tour in the place where Lewis and Clark began theirs: Louisville, Kentucky. In October of 1803, Lewis and Clark met to discuss and finalize plans for their exciting three year journey to the Pacific Ocean and back. A statue of George Rogers Clark stands at Fifth and Main Streets to commemorate this historic event. It's on Louisville's Belvedere and overlooks the Ohio River.

Louisville is a busy town with an array of activities including museums, baseball parks, and river excursions. The city also includes some of the best places to slow down with one of the most historic green spaces in the nation. In 1891, Frederick Law Olmsted was commissioned to design a park system for Louisville. He was already well known for designing New York City's Central Park and would soon be asked to design the plantings for the Chicago World's Fair of 1893. Many people are not aware of his Louisville contributions including Cherokee, Shawnee, and Seneca Parks.

Downtown, at Tenth Street between Main and Market Streets, you'll find the first stop on this tour, Old 502 Winery.

OLD 502 WINERY · *Louisville*
BROAD RUN VINEYARDS & WINERY · *Louisville*
BROOKS HILL WINERY · *Brooks*
MILLANOVA WINERY · *Mt. Washington*
WIGHT-MEYER VINEYARD & WINERY · *Shepherdsville*
FOREST EDGE WINERY · *Shepherdsville*

OLD 502 WINERY
LOUISVILLE

"Drinking wine is an experience in and of itself."
– Jon Ryan Neace –

Driving through downtown Louisville, visitors may catch glimpses of Louisville's Slugger Field, KFC Yum Center, The Muhammad Ali Center, Frazier Arms Museum, and The Kentucky Science Center, just to name a few. Any of these are worth the trip, but the Old 502 Winery is a must see. Just a short walk from the Ohio River, it's the perfect place to stroll with a glass of wine and take in some of Louisville's culture and history.

This winery describes itself as "Louisville's only urban winery." Urban is denoted as the people and spaces within a city and Old 502 definitely captures this feeling in their chic tasting room. Hip and fun, 502 draws on the energy of the revitalized areas of downtown Louisville such as NuLu and the updated Whiskey Row. The experience at this winery is enhanced by the area, but doesn't take itself too seriously. This comes through in the names and descriptions of some of their wines including "Boredough, spelled accordingly," "Sounds like Shar-Duh-Nay, tastes like wine" and "Bach's Wine" with the tag line, "It's music to your lips." Though 502 is relaxing and unpretentious, winemaker Logan Leet's mission is to make outstanding wine. They are each crafted with care and caution, creating a flavor sensation in every bottle.

Owner Jon Ryan Neace has a vision that will offer new and exciting happenings for years to come. He plans to develop 5 or 6 premier wines and compete in international competitions

Old 502 Winery
120 South 10th Street
Louisville, KY 40202
502-540-5650

Events: Yes, indoor seating for up
to 150.

to expand the Old 502 Winery name. A wine club is also a new venture and Mr. Neace is very interested in expanding the wine industry in Kentucky. "At Old 502 Winery, we aim to create top tier quality wines made from grapes grown by local, Kentucky farmers crafted in historic Downtown Louisville and delivered to you at a price that fits everyone's budget. We hope you enjoy drinking it as much as we enjoy making it for you!"

Although relatively new, 502 Winery is housed in a historic building dating from the 1880s. A picture after Louisville's 1937 flood shows that a pharmacy named Luckert's did business out of the building and also sold "Whiskey at Cut-Rate Prices."

This winery is the perfect blending of what's old and new in Kentucky's wine making industry. Stop in and sip a glass, or bottle, soon.

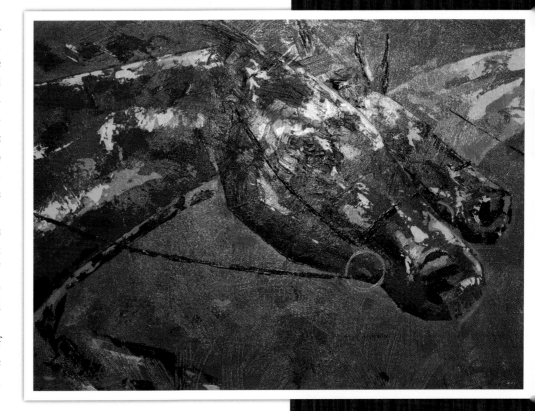

Wines

☙ B*%#@!n Barrel Red

They call this "The wine that made us famous" and we can see why. Aged in barrels that once held bourbon, this wine has a wonderfully, unexpected twist of flavor. Order your favorite pizza and start sipping.

☙ Shar-Duh-Nay

This un-oaked white has won many awards for its fresh, citrus and pear flavor. Great with guacamole and/or garlic, pork, and curry dishes.

☙ Boredough

A blended semi-dry red. This wine pairs well with red pasta sauces and beef dishes.

☙ Bach's Wine

A sweet red that's perfect to take to an outdoor concert. Be sure to bring some fruit, strong flavored cheeses such as goat or bleu cheese, and chocolate.

☙ Kentucky Lady

A sweet and fruity blend that's perfect for sipping. Enjoy with ham, deviled eggs, garlic pasta salad, hey, I think we have a picnic.

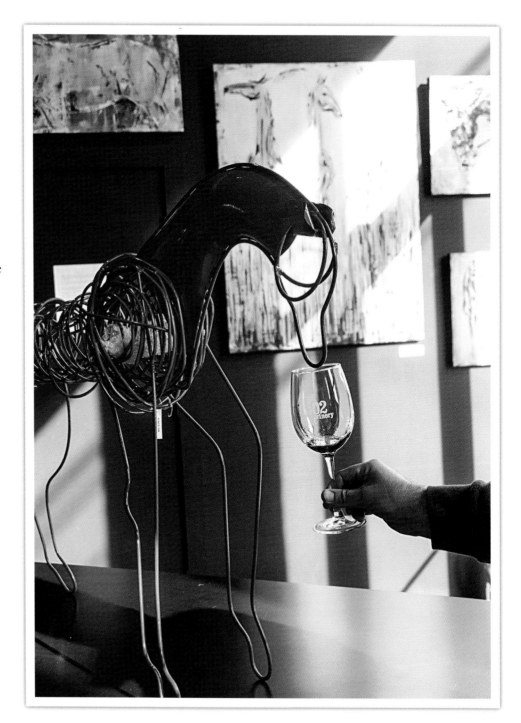

Recipe

Urban Cowgirl Chili

1 cup Old 502 Winery B*%#@!n Barrel Red Wine

1 lb. pork roast, cut into ½ inch cubes (most grocery stores will do this for free)

1 large onion, finely chopped

1 celery rib, finely chopped

2 Tbls. canola oil

3 garlic cloves, minced

1 can chopped green chilies

2 packets of chili powder

32 oz. beef broth

3 lb. boneless beef chuck roast

6-oz. can Italian style tomato paste

28-oz. can diced tomatoes

1 can black beans, drained and rinsed

1 can kidney beans

1 can hot chili beans

2 Tbls. grated dark chocolate or more!

3 tsp. ground cumin

½ tsp. salt

1 tsp. ground mustard

½ tsp. cayenne pepper

In a crock pot, cook meats for 6 hours or until cooked and can be easily shredded. Drain and set aside. In a Dutch oven, sauté onion and celery in oil until crisp and tender. Add the garlic, chilies and chili powder and cook for 1 additional minute. Stir in broth, wine, tomato paste, tomatoes, beans, chocolate, cumin, salt, mustard and cayenne. Add meat and bring to a boil, reduce heat and simmer for 30 minutes. This is fun to fix and later take to a campfire!

This recipe is a cowgirl's dream. Chocolate and wine!

73

BROAD RUN VINEYARDS & WINERY
LOUISVILLE

"If you want to have good wine, you have to grow good grapes."
– Marilyn Kushner –

The scenery in the southern part of Jefferson County moves quickly from urban to rural with large stretches of land between houses. Highways rapidly lead to small back roads running beside rock strewn creeks. Upon meeting the matriarch of Broad Run Winery, Marilyn Kushner, visitors are impressed with her knowledge of the entire wine-making process from planting to bottle. To say she and her husband Jerry take wine seriously, is a grave understatement.

"After we bought the land, we planted grapes before we even broke ground on the house," Marilyn explained. They believe in nurturing high quality grapes. "We planted 15 varieties just to see what would grow here. To our surprise, most of them did!" Their first vintage, produced in 1988, prompted a wine connoisseur to remark, "This is of commercial quality." They applied for a commercial license in 1992 and presented their first wines to the public in 1994. All of their grape wines are developed from grapes grown on the estate. For the strawberry and blackberry wines, they purchase Kentucky fruit and they proudly serve Kentucky made products.

Visitors are encouraged to bring their own food, buy a bottle or two, and settle themselves on the large back deck, which affords views of a pond and the abundant grape vines that seem to stretch for miles.

Broad Run Vineyards & Winery
10601 Broad Run Road
Louisville, KY 40299
502-231-0372

Events: Yes, indoor seating for up to
   250, outdoor seating for up to 250.

People also wander the vineyards, pet the horses, and explore the beautiful property of the Kushners.

The care taken to grow quality grapes shines through from the first sip to the last. Coupled with this is exceptional customer service. Visitors take a sitting tour of the wine and vineyards during tastings, which take at least an hour. Thoroughly trained personnel travel through history while describing the reasoning behind the color, bouquet, and taste devoted to each wine. Their wines are as close to natural Kentucky as can be found in any winery in the state.

Along with the wine education visitors receive, Broad Run Winery offers wildflower walks, concerts, and other special events throughout the year. Plan a trip soon to savor an afternoon with the Kushner's and their wonderful wines.

Wines

❧ Riesling

Considered "The epitome of Riesling," meaning a pure, white wine that expresses the territory in which it was grown, this grape originated in Germany's Rhine region. Pair with spicy Thai dishes and curries.

❧ Gewürztraminer

This wine has an exotic flare which is intoxicating to the senses. It pairs well with ham and cheeses such as gruyere.

❧ Carmine

This wine is only being produced in Kentucky at Broad Run Winery. Aged in stainless steel, it will enhance the flavor of well-seasoned beef and sausage dishes.

❧ Gala

This delightful concoction is made from Kentucky apples. It goes well with a medium or sharp cheddar and fresh fruit.

❧ Rosy Future

This is an off-dry rosé wine. Rosé gets its name and unique color from grape skins left in the juice until it reaches a slightly pinkish color. This is the wine to pair with a backyard barbecue as it goes with burgers, ribs, ham, and even pizza.

Recipe

Carmine Ragusa Spaghetti Sauce
(The Big Ragoo)

1 ½ cups Broad Run Winery's Carmine Wine

4 cans diced tomatoes with garlic, oregano, and basil

1 lb. sweet Italian sausage (Because Carmine Ragusa was a sweet Italian)

16 oz. spaghetti, cooked and drained

½ cup freshly grated parmesan-Ramona cheese

Pour wine and tomatoes into large skillet. Cook on medium-high, smashing tomatoes with fork, until desired thickness, about 20 to 30 minutes. Meanwhile, brown Italian sausage in skillet, drain and add to sauce. Add spaghetti to sauce and put on large platter. Sprinkle parmesan/Romano cheese over top. Serve hot with side salad and crusty bread. Pair with Carmine Wine.

This sauce gets its name from both
Broad Run Winery's Carmine wine
and the beloved character Carmine Ragusa from
TV's Laverne & Shirley.
Both are robust and full of life.
We think they complement each other very well.

BROOKS HILL WINERY
BROOKS

"The little winery up the hill."
– Mike Hatzell –

Once drivers leave 65, go over a set of railroad tracks, and quickly find themselves climbing a steep, curvy hill that doesn't seem to end, they will know they are almost to Brooks Hill Winery. Be sure to slow down when you spot the grape vines on the right, it's easy to pass up this little gem. The owners of Brooks Hill Winery are listed as "Mike, Karen, and Lili Hatzell." This might not sound strange, until you realize Lili is the winery dog, a lovely black lab/poodle mix known as a Labradoodle. Their bestselling wine, also carries this calm dog's name—Lili's Red. If visiting around January 26th, visitors may be treated to cake in honor of Lili's birthday.

Owner Mike Hatzell discussed his wine path one hot summer day while sitting under the large pavilion behind the winery. Hatzell was first introduced to vineyards during an extended stay in France during 1964 "at the government's behest." In other words, he was drafted into the Army. While in the military, he would take long drives into the French countryside and became intrigued by the acres of grapes he saw growing there. He wanted to taste this famous wine in the place it was made. "The first time I tasted it, I spit it out." While becoming accustomed to the taste of wine, the palate must be trained. This happens by starting with the sweet wines, later moving to the semi-dry/semi-sweets, and then developing a taste for the dry. Mr. Hatzell started with the dry, thus his adverse reaction.

Brooks Hill Winery
2746 Brooks Hill Road
Brooks, KY 40109
502-957-7810

Events: Yes, indoor seating for up to
50, outdoor seating for up to 200.

After Hatzell's semi-retirement, he and his wife opened the winery in 2007. A nice variety of berry and grape wines can be found on the shelves and customer input is what keeps Mr. Hatzell trying new combinations. "Our Black & Blue is the newest wine, and it's selling so well that it just might pass Lili's Red as our best seller." Black & Blue is a mixture of their blackberry and blueberry wines.

They grow some of the grapes used in their wines on the property, and buy the remainder from Kentucky farmers. For fruit wines, however, Mr. Hatzell has an admirable philosophy. "We go where we can get the best fruit." Marion blackberries from Oregon have a reputation for being the best, so that's what Mr. Hatzell uses in his blackberry wines. With their inside and outside seating areas and free entertainment every other weekend from spring to fall, a trip to Brooks Hill is well worth the drive to "the little winery up the hill."

Wines

❧ Coral Ridge Red

A bold red wine crafted from a blend of Cabernet Sauvignon, Merlot and Cabernet Franc grapes. Pairs well with barbecue, lamb chops, and tomato based stews.

❧ Morning Sky

A dry, white wine from 100% Kentucky Traminette and Vignoles grapes tastes of citrus, honey, peach and dried apricots. This wine pairs well with poultry, pork, and fish.

❧ Rose of Sharon

The fruitiness, sweetness, and crisp acidity of this wine make it a perfect match for lighter dishes, such as fruit and vegetable salads, or enjoyed chilled on its own.

❧ Lili's Red

Named for the winery dog, this well-balanced wine is made from the highly aromatic Concord grape. Wonderful alone or with cheesecake or chocolate covered cherries.

❧ Chocolate Razz

A port-style dessert wine, the blend of raspberry wine and chocolate is so delicious you'll want to pour it over your pancakes, ice cream, cheesecake, etc…or just sip it!

Recipe

Lazy Blueberry Wine Cobbler

½ cup Brooks Hill Blueberry Wine
1 cup butter
1 cup sugar
1 cup self-rising flour
¾ cup milk
1 can blueberry pie filling

Preheat oven to 350°. Place butter in 9x13 pan and set in oven to melt. In small bowl, mix wine and pie filling. Set aside. In medium bowl, mix all other ingredients thoroughly. Remove pan from oven and pour flour mixture over melted butter. Pour blueberry-wine mixture over top of batter. Do not stir. Bake until the top is browned, about 25-30 minutes. Serve with vanilla ice cream. Mmm. a sweet summer treat any time of the year.

MILLANOVA WINERY
MT. WASHINGTON

"A Family Tradition."
– John Miller –

MillaNova sits deep in the heart of Bullitt County, and Bullitt County residents. It was voted the county's best winery several years in a row. This winery is close to both Louisville with all its things to do, and Bardstown with all its simplistic pleasures.

The name of the winery is a combination of owners John and Donna Miller and her father Frank Terranova.

When he was barely 20 years old, Donna's grandfather left Sicily and came to Americ by way of Ellis Island. He passed the Italian harvest-time ritual of home winemaking to his children, who passed it on to theirs. The Millers kicked around the idea of opening a winery for years before they took the plunge. "I always liked wine and made it as an amateur on a trial and error basis," Mr. Miller said. When he received enough positive reviews from friends and won a few amateur wine-making awards, The Millers decided to open the winery. Although Mr. Miller still considers himself employed as a builder and developer, he's very much hands on at the winery. "We decide which wines to develop and keep based on my tastes, along with customer feedback." He proudly spoke of certain vintages as if they were his children. "First Lady of Kentucky, Jane Beshear, chose the Cabernet Sauvignon for the inaugural ball. She liked it so well she put it in every Kentucky State Park that serves alcohol. We've won awards on every wine that we make. MillaNova is the only winery in

MillaNova Winery
744 Gentry Lane
Mt. Washington, KY 40047
502-664-8304

Events: Yes, indoor seating for up to 250, outdoor seating for up to 300.

the country making a walnut wine." John Miller's not bragging, just proud of how far the winery's name has grown in such a short time.

In addition to the wines themselves, MillaNova has become well known for the events they host. Weddings are held many weekends and the Millers created relaxing indoor and outdoor spaces in which to enjoy the landscape. There is a large pavilion where entertainment is offered. A recently added feature is a fire pit area with seating. Although they serve a cheese plate using Kentucky Proud cheeses, they encourage visitors to bring their own food, buy a bottle and enjoy the atmosphere.

The concerts held throughout the summer are casual affairs attended by many. Most visitors know to bring their own chair for the concerts, as seating space fills up quickly.

The Millers invite you to visit their winery soon and "soak in the atmosphere." You'll be glad you came.

Wines

❧ **Sweet Falls of Niagara**

A heady, sweet white bursting with intense grape flavors and floral aromas. Enjoy this one chilled while sitting in a hot tub or inner tube. Refreshing!

❧ **Mint Julia**

This wine is reminiscent of the traditional Kentucky Mint Julep. It is best served over ice with a sprig of mint.

❧ **Alyssa's Blush**

A sweet, well-balanced blend of the Cayuga, Niagara, and Concord grape with robust fruit flavors.

❧ **Chardonberry**

MillaNova's own special blend of Chardonnay with a trace of strawberry for a refreshing semi-sweet blend.

❧ **Strawberry Blonde**

A big, intense wine; like fruity strawberry jam with a kick.

Recipe

Jalapeño Stuffed Mushrooms

¼ cup MillaNova dry Reisling

18 large mushrooms tops, boiled 5 minutes and drained well

8 oz. cream cheese, softened

1 jalapeño, chopped fine, seeds and ribs removed (The seeds and white rib inside peppers contain the heat of the pepper. If you prefer the recipe spicy, leave some or all of them in.)

3 slices of bacon, cooked and crumbled

½ cup shredded sharp Cheddar cheese

Squeeze mushrooms between paper towels to remove excess water. Mix wine, cheeses, jalapeño pepper, and bacon, together thoroughly. Fill mushrooms with 1 teaspoon of filling and bake at 350° 10-15 minutes until cheese is hot and begins to melt. Serve immediately.

"What do you call a nosy pepper? Jalapeño business."

Everyone will be asking for the secret to this recipe which combines two favorite appetizers, stuffed mushrooms and jalapeño poppers.

"A complete sensory experience awaits."
– Jim Wight –

The gravel drive leading visitors to Wight-Meyer Winery is reminiscent of days gone by before many roads were concrete or black top. Driving up this path on a summer evening, music drifts over the vineyards, drawing people toward tantalizing scents, wonderful wines, and a big sense of camaraderie. Owners Sandy and Jim Wight honored her father, Herbert Meyer, when naming the winery. He once owned the land where Wight-Meyer Winery now stands. They do his name proud by carrying on the agricultural tradition he started with his family. Jim began growing grapes alongside his brother-in-law, Butch Meyer, a vintner well known in the wine community. Wight tried many varieties of grapes and blended them. They were so well received he began entering amateur competitions. "After I won so many amateur medals," Jim Wight said. "Sandy convinced me we should open a winery." He joked that she just wanted somewhere to hang the medals other than her kitchen. Becoming a winery owner has served him well as he is the vintner for his and other wineries, plus a consultant for many others. In addition, he is a past president of the Kentucky Vineyard Society. While in the post, he worked with the University of Kentucky's agriculture department to plant Moore's Diamond Grape. He uses this grape to make the wine he calls, "the one of which I'm most proud," Kentucky Diamond.

Wight-Meyer Vineyard & Winery
340 Meyer Drive
Shepherdsville, KY 40165
502-921-0267

Events: Yes, indoor seating for up to
130, outdoor seating for up to 260.

Mr. Wight enjoys pitting his wines against others' in competitions, including the Indianapolis International Wine Competition. "The Winter Solstice special release ice wine has won best of class in Indianapolis three years in a row."

Although this picturesque setting has an array of wines for any palate, what visitors enjoy the most about the winery may surprise some; it's Henry the dog. A gentle black Labrador, he makes the rounds between tables and visitors as if he's personally greeting each and every person.

Jim Wight likes to discuss some favorite recipes and wine pairings. "I have a spicy jerk chicken sandwich that pairs well with our Chambourcin. It also goes well with salami and pepperoni, any extra flavorful meat." Wight-Meyer hosts music every other Saturday from April through November. Bring a chair, just in case the ones provided are filled, and listen to some great bands under the stars while sipping some of Jim Wight's wonderful choices.

Wines

❧ Petit Syrah

A deep blackberry flavor with vanilla tones. Serve this with a seasoned burger and sweet potato fries.

❧ Chambourcin

Notes of blackberries and cherries with a hint of oak. Very good with spicy meats such as salami and pepperoni.

❧ Pine Creek Summer

Fresh tropical fruit flavors with a smooth citrus finish. Just right chilled on a hot summer night down by the creek.

❧ Vignoles

The crisp taste of apricot and pear with a citrus finish. Pair with salmon, a Swiss cheese plate, or a fruit salad.

❧ American Diamond

A burst of fruity flavor, crisp and refreshing. Diamond is a double-gold winner. The perfect wine to serve with chicken salad, crab cakes, or pork chops.

Recipe

Polynesian Meatloaf for a Crowd

1 and 1 ½ cups Wight-Meyer Chambourcin, divided
2 lbs. lean ground beef
1 lb. mild, medium, or hot sausage, according to taste
1 cup quick oatmeal
1 small, sweet onion, diced
1 red bell pepper, diced, stem and seeds removed
2 cloves garlic, diced
3 Tbls. olive oil
1 can pineapple rings
10-oz. jar sweet and sour or duck sauce

In a large bowl, mix together ground beef and sausage. In skillet over med-high heat, cook onion, pepper, and garlic in oil until tender and slightly browned. Add to ground beef/sausage mixture, pour 1 cup of wine over all, add oatmeal, and mix well. Form into 10 small meatloaves and place in a single layer in large 9x13 pan. Mix together sweet and sour sauce and remaining ½ cup of wine. Put one pineapple ring on each meatloaf. Pour sweet and sour/wine mixture over all. Bake at 350° for 35-40 minutes. Tasty! We love this recipe because we can make it up early in the day, slide it in the fridge ready to go in the oven, then pull it out and bake when needed. When served on hamburger buns, this dish goes over especially well with hungry teenagers.

"In water one sees one's own face;
but in wine one
beholds the heart of another."
– French Proverb –

FOREST EDGE WINERY
SHEPHERDSVILLE

"We solve the world's problems one glass at a time."
– Brance Gould –

Almost directly across from this winery sits Bernheim Forest. It is a nationally recognized arboretum with picnic areas, fishing lakes, and hiking trails. About a half mile down the street is Jim Beams Outpost, which offers distillery tours and bourbon tastings. These attractions, plus the winery, make this area a perfect place to visit and explore the beauty, and libations, which enhance Kentucky's pleasures.

The owners of Forest Edge Winery, Traci and Brance Gould, are excited about the opportunities the business affords them. The wine, Zoey Rosé, is named after a family member born with only one malformed kidney. One dollar from each bottle is donated to The Kidney Foundation. In addition, Forest Edge hosts the Zoeyfest every June to help raise awareness of the foundation and all their work. "We're passionate about what we do because we love wine and love the business. It allows us to give back," Brance explained.

Mr. Gould is excited to see the growth in the wine industry across the state. He especially likes the appreciation shown to it by young adult drinkers. "I really like seeing people in their mid-20s that have a wine palate--an appreciation for wines." His appreciation came through his father-in-law, Butch Meyer. That's a name well known in Kentucky wine as he is the vintner for several wineries and the consultant for many others. Mr. Meyer has introduced countless people to the love of wine-making.

Forest Edge Winery
1910 Clermont Road
Shepherdsville, KY 40165
502-531-9610
855-355-9463

Events: Yes, indoor seating for up
  to 50.

Not all of them opened wineries, but many have gone on to grow grapes, produce wines, and win medals in amateur competitions.

Mr. Gould's vineyard path started about 2007 when he entered his first wine, an ice-wine made in honor of his wife, in the Kentucky State Fair competition for amateur vintners. It won a silver medal and gained the wine-making industry another patron. Mr. Gould began looking for a building, close to the interstate, to open a winery. He found one, negotiated for about 6 months, renovated for another 6, and then opened in 2011.

"When people walk into the tasting room, they often comment on how pretty the space is, and that's the reaction I wanted." A sign announces, "Enter as strangers, leave as friends." In addition to the beautiful space, Forest Edge offers something few wineries do, a kid's room with a TV, Legos, and a coloring station.

A visit to Forest Edge is the perfect way to cap off a superb day of Kentucky sightseeing. Come visit the Gould's soon and make some new friends.

Wines

- **Black Cherry Pomegranate**

  An interesting blend with a tart finish. One of our favorites.

- **Zoey Rosé**

  A blend of Niagara and Concord grapes. Forest Edge recommends that it be "served chilled and shared with friends." A cheese plate would go well too.

- **Chocca-Con**

  One of Forest Edge's bestsellers. Concord infused with dark chocolate, they suggest you pair it with chocolate to enhance the flavor.

- **Love Spell**

  The ice wine owner Brance Gould developed for his wife Traci. Diamonds are a girl's best friend, as this wine from the diamond grape proves. Pair with a fresh peach cobbler, or any fruit pie.

While we teach our children all about life, our children teach us what life is all about.
— Angela Schwindt

Recipe

Wine-Doughs

2 cups Forest Edge Rendezvous

3 egg yolks, beaten

1 tsp. baking soda

About 10 cups of self-rising flour, more or less if needed

About 16 ounces warm honey

Powdered sugar, for dusting

2 cups canola oil

This recipe can easily be halved. In heavy-bottomed 4-6 cup sauce pan, heat oil to between 360° to 375°. Mix all other ingredients except honey and powdered sugar. If batter is still sticky, add enough flour to make a stiff dough. Roll dough to ½ inch thickness on floured surface. Cut with a small glass, biscuit, or cookie cutter. Deep fry, turning with tongs to ensure they're evenly browned. Dip in warm honey and sprinkle with powdered sugar. Best served warm. Just one won't do, for these doughnuts have two.

*Served warm
or cold,
these yummy
doughnuts are a
crowd pleaser!*

Central Kentucky Region

The Old
TALBOTT TAVERN

1779

1779

BARDSTOWN AREA
LEBANON • BARDSTOWN • BLOOMFIELD • WILLISBURG

The History Tour

Throughout this area of central Kentucky, historic markers about important people and events in America's past can be seen in many places. Bardstown, just one of many beautiful small towns in central Kentucky, contains so much of the country's history, visitors could spend days here and not see it all. Luckily, the Bardstown/Nelson County Historical Museum sits at 114 North Fifth Street, for those wishing to study the town and county a little more closely.

The Old Talbott Tavern, once a stage coach stop, boasts a visit from Abraham Lincoln, a few bullet holes, and ghosts. Many employees, locals, and visitors have told stories of ghostly encounters. Some believe that one ghost is Jesse James while another is known as the Lady in White. On Friday evenings at nearby Wickland, home to three Kentucky governors, there's a family-friendly paranormal experience with a psychic medium. Reservations are required and these tend to be very popular.

Hurst Drug Store and Soda Fountain has been a fixture on Bardstown's square since the 1940s. While sitting at the wide marble counter ordering a hamburger and soda, visitors might look around expecting to see poodle skirts and blue suede shoes.

My Old Kentucky Home State Park is the site of Federal Hill, the home Stephen Foster celebrated in song. My Old Kentucky Home became the official state song in 1928. Every year, a popular outdoor musical, The Stephen Foster Story, commemorates Foster's love of the home and Kentucky.

Old Bardstown Village re-creates a 1790s' frontier village and is a representation of the first westward expansion movement in America. Sitting next door is the Civil War Museum and the mansion where the first Confederate Flag was flown in Kentucky.

There are many bourbon distilleries in Bardstown, including Heaven Hill and Four Roses. Ninety-five percent of all bourbon whiskey produced, comes from Kentucky.

Into the midst of historic Bardstown, several wineries have arrived and set up shop. These wineries enhance the quaint atmosphere of the region, while offering another Kentucky made delight.

WHITEMOON WINERY · *Lebanon*

MCINTYRE'S WINERY & BERRIES · *Bardstown*

CHUCKLEBERRY FARM & WINERY · *Bloomfield*

SPRINGHILL WINERY

PLANTATION BED & BREAKFAST · *Bloomfield*

HORSESHOE BEND VINEYARDS & WINERY · *Willisburg*

WHITEMOON WINERY
LEBANON

"Wine makes life better!"
– Alexandra Ackerman –

Although Marion County has only about 20,000 residents, the local WhiteMoon Winery is introducing the world to a slice of Kentucky life.

Canton Cooperage, a barrel making business nearby, supplies barrels all over the world. "We hosted Canton Cooperage's international business meeting," said tasting room manager Michele Espinosa. "There were people here from Spain, Italy, France, Argentina, China, Australia, and Japan, just to name a few. We held a big cook-out for them and they loved it. Grilling is such an American thing and they wanted to experience as much of us as they could." They loved their time in Kentucky and all hoped to return soon.

Canton barrels not only hold wine for aging, but earned a prominent spot on Mooned Red Wine when Simple Pleasures Vineyard owner, Jimmy Hatchet, agreed to undress and wear a Canton Barrel on the label. Upon seeing this label, visitors can feel the fun-loving atmosphere at this winery. The photograph is portrayed in black and white, reminiscent of the good old days when hard work and waving to passing cars were the norm. The nostalgia at WhiteMoon feels as if it's part of the landscape.

WhiteMoon may be one of the newest wineries on the block, opening in 2013, but the vintner, Lisa Wicker, has a lot of experience making wine. Since she's been involved with wine for most of her life, Wicker is very particular about the ingredients that go into her wines. "I've hand carried yeast on the plane back from California when it's too hot to ship." The heat can

WhiteMoon Winery
25 Arthur Mattingly Road
Lebanon, KY 40033
270-402-1285

Events: Yes, outdoor seating for up to 100.

affect the ingredients, just as it does the wine. Wicker does not recommend leaving wine in a hot car, even for a short time. She likes to go into the fields where WhiteMoon purchases their grapes and test for ripeness. "Picking too early or leaving on the vine too long can both affect the taste." Wicker likes to pick them all quickly and get them into the vats before the grapes lose any flavor.

Owner Alexandra Ackerman acquired the property about ten years ago, but only began experimenting with wine about 6 years ago. She wanted to honor the land's former owners by keeping it agricultural. She had the soil analyzed to see what could grow successfully and the results showed grapes as a possibility. Though the original farmhouse is gone, Ackerman is trying to preserve two large stone fireplaces on the property. The plan is to restore them and host musical groups outside on summer and fall evenings. Although this may be one of Kentucky's newest wineries, it won't be long before it will be the favorite of many across the bluegrass, and beyond.

Wines

❧ **Chambourcin**

Smooth with a slight spice, this excellent wine would pair well with a favorite beef roast, salami, pepperoni pizza or mac and cheese.

❧ **Mooned Red**

This sweet table wine is great to sip by itself or with hard cheeses, roasted vegetables, or pork dishes.

❧ **Traminette**

Fruity and semi-sweet, this wine is perfect with a blackberry cobbler, cured meats, or a tray of soft cheeses.

❧ **Chardonnel**

The grape flavor bursts into the mouth with this delicious white wine. Serve with French baguettes and a tray of spreadable cheeses.

Recipe

S-Wine Smoked Sausage Corn Chowder

1 cup WhiteMoon Traminette
24 to 30 oz. smoked sausage
64 oz. vegetable broth, divided
1 ½ to 2 cups sliced green onions
4 tsp. vegetable oil
28 oz. potatoes O'Brien
28 oz. shredded hash browns
Two 13 to 15-oz. cans creamed style corn
⅛ tsp. ground red pepper
1 cup heavy cream

Slice sausage into ½-inch thick slices and simmer in 32 oz. broth and Traminette. In a tall stock pot, cook onions in oil over medium heat for 3 to 4 minutes or until tender. Stir in sausage and wine mixture, remaining broth, potatoes, corn and red pepper. Bring the chowder to a boil and let simmer, uncovered, for 20 minutes. Stir in cream, heat through, garnish with additional green onions if desired and serve. You can also make the chowder the night before, leave out the cream and onions for garnish, refrigerate and put it in a crock pot in the morning. Add the cream and additional onions when you are ready to eat.

"When you ask one friend to dine,
give them your best wine!
When you ask two, the second best will do!"
– Henry Wadsworth Longfellow –

MCINTYRE'S WINERY & BERRIES
BARDSTOWN

"I grow my own country wine."
– Tommy McIntyre –

Sitting just outside of Bardstown is a small winery that many may overlook as they drive past. McIntyre's Winery and Berries is an unassuming building which reflects the humbleness and charm of this winery. Owner Tommy McIntyre is one of

the most generous people you're ever likely to meet. When we arrived on harvest day, the blueberries were all picked and the blackberries were a few days away from ready. He graciously offered us lunch, gave us blueberries, and pretended not to notice when Kathy got carried away with the ripe blackberries. He told us his story and the personal reasons that led Mr. McIntyre down his winery path.

"My grandparents used to own the land the winery sits on," he said. "I stayed with my grandmother a lot as a child. We had no electricity and no running water. We cut wood for supper and shot squirrels to eat." Although a hard life, like many Kentuckians who choose the agricultural lifestyle, Tommy McIntyre loved it. When there's no electricity, families naturally spend more time together, talk, read, and listen to each other.

It was a relative who first introduced him to winemaking. "An older cousin gave me my first sip of homemade wine." McIntyre wanted to learn more about this concoction and began making his own from wild berries. They became harder to find, so he grew his own. Mr. McIntyre may not have ever opened a winery, were it not for the economy. "The company I worked for got bought out and I lost my job." He and his wife Debbie already owned the land and decided to open a u-pick berry farm. It wasn't long before his wine curiosity returned and McIntyre began experimenting with the berries to see which made

McIntyre's Winery & Berries
531 McIntyre Lane
Bardstown, KY 40004
502-507-3264

Events: No

good wines. He discovered most of them did and tried something very unique with his berries.

"I make a dry, semi-sweet, and sweet of each of the wines I make. I haven't seen anyone try it and I thought I would. People seem to like it." Yes they do. His black wine is a dry blackberry aged in black rum barrels and the taste is tannic and interesting. His blueberry in the semi-dry could be added to any fruit cobbler, pie, or syrup to enhance the flavor. In addition to the blackberry and blueberry wines, peach, elderberry, pear, and strawberry wines are offered, with more flavors planned in the future. Mr. McIntyre's winery is different from a lot of others for two reasons; he works hard to grow or procure all the fruit as close to his farm as possible and he's willing to experiment with the sugar contents of the wine. There is no such thing as the complete right way when it comes to making wine. As long as people like it and request it, it's the right way. The next time you're in the Bardstown area, be sure to stop by, say hello to Mr. McIntyre, and taste some of his fantastic experiments.

Wines

🍃 **Blackberry Wine**

Grown from fruit on Mr. McIntyre's property, this wine comes in sweet, semi-sweet, and dry varieties.

🍃 **Blueberry Wine**

Estate grown, this customer favorite is available in sweet, semi-sweet, and dry.

🍃 **Black Wine**

Blackberry wine aged in black rum barrels for a very different, interesting flavor.

🍃 **Elderberry Table Wine**

This taste of pure Kentucky is aged in stainless steel. It's smooth and mellow flavor is best sipped slightly chilled.

Recipe

Not My Grandmother's Jam Cake

1 cup McIntyre's Blackberry Wine
Spice cake mix (2 layer size)
⅓ cup water
3 eggs
18-oz. jar blackberry jam, divided
1-2 tablespoons of flour, for dredging
1 cup chopped walnuts and/or raisins, if desired
8-oz container of whipped topping
12-oz. jar caramel ice cream topping
Blackberries, for garnish

Preheat oven to 350°. Grease and flour three 8 or 9-inch round cake pans. Combine cake mix, wine, water, eggs, and ½ of the jam. Beat on medium speed for 1 minute, then high for 1 more minute. Dredge walnuts and raisins with flour and fold into batter. Bake for 25-40 minutes, depending on the thickness of the cake pans. Cool completely. Place one layer of cake on plate or cake stand. Poke at least eight holes in cake with the end of a wooden spoon. Drizzle top with ⅓ of the caramel, reserving the rest for remaining layers. Mix remaining blackberry jam with whipped topping. Put two dollops on top of caramel topping and spread evenly. Repeat for the remaining two layers, garnish with blackberries.

Sorry grandma. Your 50 ingredient, 3-hour Jam Cake, is the best in the world. However, there are friends to enjoy and wines to drink, so I mimicked your recipe. I think you would approve, except for the cake mix part.

"Get your Chuckle on!"
– Chuck and Ladonna Hall –

Sitting just outside Bardstown's city limits is a dream become reality for some hard-working Kentuckians. Chuck and Ladonna Hall met on a blind date while in high school. They married in 1991 and started working toward their dream of farm ownership. As he learned the construction business, eventually starting his own company, she worked as a waitress and at a law firm while earning her paralegal degree from the University of Louisville. At the same time, they purchased some cows and leased tobacco base to earn extra money. Within a year of their marriage, they had their first of three girls, and began handing down family traditions.

By the time their third daughter was born, Chuck had begun playing around with making homemade wine. In 2003, they purchased their dream land, thirteen acres on Garrison Lane in Bloomfield. Together, they decided to specialize in berry wines and planted blackberries, blueberries, strawberries, and raspberries on their farm and opened the winery on May 7, 2010. Readers of Kentucky Living Magazine voted them Best Kentucky Winery of 2013.

According to Chuckleberry's website, when it came to naming the winery, the Hall's chose a nickname Chuck's mother, Doris, used to call him as a child. Unfortunately, she passed away a few years ago and never got to see her son and daughter-in-law's winery dream come true. "She would have lived here

Chuckleberry Farm & Winery
527 Garrison Lane
Bloomfield, KY 40008
502-249-1051

Events: Yes, indoor seating for up to
150, outdoor seating for up to 150.

and been such a huge part of it," Chuck said. "We decided to keep her involved and name it after her loving nickname for me."

The picturesque scenery of this farm winery includes a 2-acre lake with a pavilion where many wedding ceremonies and photography sessions are held. There are a lot of green spaces to spread out a blanket and enjoy the sunshine, lunch, and a bottle of wine with that special someone, or several special someones.

In addition to wine, Chuckleberry's has a commercial kitchen on site where they make homemade jams and jellies and their famous wine fudge. Honey, fresh and raw from their farm, is also available. It's been reported in many sources that a little local, raw honey taken daily will help people build their immunity to plants that cause seasonal allergies. Pregnant women should not eat raw honey and children under one year of age should not eat honey of any kind.

Another service offered at Chuckleberry's is customized labels on all their products for special events. The Hall's and their daughters work hard to make every visit an experience. The good people at Chuckleberry's Farm Winery would love to share their "little piece of Heaven" with you soon.

105

Wines

❧ **Sweet Cherry Wine**

Delectable and refreshing, this wine also comes in a "Sour Cherry Pie" variety. Sip alone or with cheesecake.

❧ **Honey Apple Wine**

Using Kentucky fruit and some of his very own honey, Chuck presents this sweet wine with a refreshingly different taste. Serve chilled or warm.

❧ **Black Raspberry**

This sweet wine is made from fruit that looks and tastes like a raspberry, but has the color of a blackberry. A sweet and delicious dessert.

❧ **Choco Raspberry**

Rich and decadent, this wine is a party in your mouth. The really should re-name it PMS wine.

Recipe

Country Wine Pear Pie

1 cup Chuckleberry Farm and Winery Pear Wine
¾ cup light brown sugar
¼ tsp. salt
¼ tsp. all spice
1 ¾ tsp. cinnamon
½ tsp. vanilla
⅓ cup sugar
2 Tbls. lime juice
5 cups ripe pears, cored, peeled and thinly sliced
Two ready-made pie crusts from the refrigerator section
3 Tbls. butter, cut up

In a large pan, mix wine, pear slices, sugars, salt, cinnamon, and lime juice. Bring to a rolling boil for two minutes. Remove from heat and stir in vanilla. Place crust in a deep dish pie plate. Pour in filling. Cut the other crust into lines to make a lattice design. Dot with butter. Bake at 350° 40-50 minutes, or until crust is browned. Let cool slightly, then cover with drizzle.

Shiny Wine Drizzle

1 cup Chuckleberry Pear Wine
2 ½ Tbls. flour
1/3 cup sugar

Combine in a small sauce pan and let simmer until slightly thickened. Using a cooking brush, spread over pie or drizzle to your desire. The leftover drizzle makes a wonderful ice cream topping. Especially if slightly warmed. Yummy!

SPRINGHILL WINERY PLANTATION BED & BREAKFAST

On the outskirts of Bardstown in Bloomfield, sits Springhill Winery Plantation Bed and Breakfast. Entering the grounds of this historic property, visitors may feel as if they've traveled back in time. The picturesque bed and breakfast was originally built from 1857-1859 by first owner John R. Jones. It is rumored

that many visitors have seen mysterious happenings while staying and attribute these incidents to a skirmish that occurred on the property during the Civil War.

According to legend, and some newspaper accounts, owner John R. Jones objected to some Confederate guerillas coming onto his property to gather provisions and refresh horses. When a captain decided he also wanted an expensive saddle, this proved too much for Jones, who retrieved a gun and shot at the soldiers, hitting the captain in the arm. Surgery techniques being what they were at the time, the captain's arm was removed. In retaliation, the other soldiers returned and killed Jones. When a Union troop traveling through the town was informed of the tragedy, they decided that two Confederate soldiers would give their lives for Jones's death. Two were chosen from a POW camp at Cave Hill Cemetery, brought to the property and killed in reprisal.

Since then, people have reported bloodstains where Jones's supposedly died and ghostly apparitions throughout the house and grounds. The history intrigued current owners Carolyn and Eddie O'Daniel. They love the challenge of

**Springhill Winery Plantation
Bed & Breakfast**
3205 Springfield Road
Bloomfield, KY 40008
502-252-9463

Events: Yes, indoor seating for up
to 50.

operating the winery and bed and breakfast, but remember the history as well.

Every Easter, the O'Daniel's gather many family members and have a "Blessing of the Vineyards" ceremony where they consecrate the land in preparation for the next growing season. Afterwards, a large luncheon is provided and planting season can begin on the three acre vineyard.

The soil in this area, loamy with magnesium and limestone deposits, is perfect for growing grapes. Generally, the amount of both sun and rain are optimal to produce some of the finest wines. The O'Daniels encourage visitors to bring some food and enjoy the grounds. They'll help you choose a bottle of wine and point out places to spread your blanket and let the world slip away as you take in the beautiful surroundings.

The Plantation Bed & Breakfast offers individual rooms to rent or the entire house is available for special occasions. The O'Daniels would love to help you plan a special party or host an event. Guests will think life doesn't get sweeter than sitting in the gazebo, watching the sun go down, and turning an ear to listen for whispers in the dark.

Wines

Springhill Bordeaux

This dry, full-bodied red wine is smooth and lush with berry flavors. It pairs well with grilled steaks and red pasta sauces.

Magnolia Mist

A blend of Vidal and Seyval Blanc, this dry white is light and fruity with a melon finish. Serve with poultry or shellfish.

Sweet Rhett

Part of the homage to *Gone with the Wind*, which also includes Sweet Scarlet, Prissy, and Plantation White, this sweet red is wonderful to sip with dark chocolate.

Harvest Mead

A traditional Celtic wine and said to be man's oldest drink, this dessert white is medium-bodied and a Kentucky harvest in a glass. A great after dinner aperitif.

Recipe

Lemon Wine Cake

1 cup Spring Hill Magnolia Mist Wine
1 package white cake mix
½ cup vegetable oil
4 eggs
8 oz. lemon gelatin, divided
1 tsp. lemon extract

Divide gelatin in half and combine it with the rest of the cake ingredients in large bowl. Beat at low speed until moistened, increase to medium speed and beat for 2 minutes. Pour batter into greased Bundt pan. Do not flour pan. Bake at 325° for 40 minutes. Wine may cause this cake to cook too fast so test for doneness occasionally. Make sure tooth pick comes out clean. Cool for 10 minutes in pan, remove to wire rack and cool completely.

Lemon Wine Glaze

¼ cup Magnolia Mist Wine
½ cup orange juice
3 Tbls. sugar
2 Tbls. lemon juice
1 cup powdered sugar
½ packet of lemon gelatin

Heat wine until just below boiling. Add the remaining ½ package lemon gelatin and sugar to melt. Add juices, stir. Add powdered sugar to desired consistency. If too thin, add powdered sugar a tablespoon at a time. If too thick, add water a tablespoon at a time. Place the cake on platter. Pour or brush glaze over top and sides and decorate with lemons. Pretty and tasty!

HORSESHOE BEND VINEYARDS & WINERY
WILLISBURG

"Serious wine from not too serious people."
– Ann, Bob, and Greg Karsner –

Evidence that Bob and Ann Karsner do not take themselves too seriously can be seen everywhere at Horseshoe Bend Winery. Whether it is their wines named "Pimpin Penguin Riesling" or "Kong's Thong Norton" or the sign-off from their son Greg on their winery website, "If our winery keeps getting national and local acclaim, I will finally be able to afford Rogaine or a permanent set of hair plugs," visitors will appreciate the Karsner's sense of humor. One thing they're serious about, however, is making quality wines.

With its 180° views of the vineyards and the land beyond, visitors feel as if they're on top of a mountain when visiting this winery. The views prompted one visitor from Minnesota to ask, "How did you ever find heaven?" This beautiful Kentucky hillside may not be where the Karsner's originally planned to set down roots and open a winery, but they are more than happy that this is where they landed.

"We always knew we wanted to retire and open a winery," Ann explained. "We just always thought we'd be in Virginia." What changed her and her husband Bob's mind was a vacation. "We were coming home and stopped at a fast-food restaurant. The young lady at the counter was just so friendly, so accommodating. We'd missed that during several hectic years in Wash-

ington D.C." One person showing our usual down-home hospitality changed the Karsner's course and set them on a winery path straight to Kentucky.

The Karsner's have a long wine history together and started making wine while dating. They even named one of their dogs after the first bottle of wine they shared together—

Horseshoe Bend Vineyards & Winery
1187 Lawson Lane
Willisburg, KY 40078
859-375-0296

Events: Yes, indoor seating for up to
20, outdoor seating for up to 20.

Scharzhofberger, Scharz for short. While searching for ways to slow down and prepare for their own winery dreams, the Karsners lived in Holland for several years, volunteered at many wineries around the state of Virginia, and started searching for that perfect piece of ground. They found it in Willisburg. They chose this spot because the limestone in the soil is similar to French soil in wine-growing regions, which is what Bob preferred.

This fast growing winery is quickly becoming known beyond the borders of Kentucky. A bottle of Kong's Thong recently appeared in the background of the show, Two Broke Girls and their wine is sold in both Ohio and Illinois. All of this fame hasn't spoiled them though as they realize customer service is what keeps people coming back. Horseshoe Bend Vineyards and Winery is one of the loveliest places in Kentucky. Bring a lawn chair or blanket and some food and experience the lovely solitude of this winery.

Wines

☙ Rose of Cabernet Franc
This dry wine, made in a traditional French-style, displays the pink color and light berry flavors associated with rosé wines. Pair with ham, Swiss cheese, and a baguette.

☙ Traminette
Similar to Gewürztraminer, this wine is reminiscent of a lemon and lime beverage. Refreshing as a cool drink on a hot day.

☙ Red Jester Chambourcin-Malbec
This dry, full-bodied blend is bold with a spicy and oak finish. Pair with salmon, lamb, and pork.

☙ Pimpin' Penguin Riesling
Semi-sweet with flavors of tropical fruit, this is the all-around wine to serve with everything from appetizers to Asian foods.

Recipe

Rice Krispies with a Twist

¼ cup Pimpin' Penguin Reisling
3 Tbls. butter
16-oz. bag marshmallows
Red food coloring
12-oz. box rice cereal
½ cup powdered sugar
Red hots or any red candy, if desired

In Dutch oven or large heavy pot, on med-high heat, bring wine to rolling boil for two minutes, reduce heat. Add butter to melt. Add marshmallows and stir until the mixture is smooth. Remove from heat. Add red food coloring a few drops at a time and stir until there is consistent color. Stir in rice cereal until completely covered with marshmallow mixture. Grease cookie sheet and press mixture onto cookie sheet until it is ½-inch thick. Tip: cover rubber spatula with oil or cooking spray so it does not stick to the mixture as it is pressed into the pan. Let stand for 30 minutes, then refrigerate for 30 minutes. Sprinkle powdered sugar on display plate. Grease your cookie cutter and cut mixture, or shape yourself, however you wish. Sprinkle again with powdered sugar and press Red Hots into the top. Enjoy!

*Snap!!!
Crackle!!!
Cheers!!!*

Central Kentucky Region

FRANKFORT AREA

SHELBYVILLE · LAWRENCEBURG · FRANKFORT · MIDWAY · LEXINGTON · VERSAILLES

The Thoroughbred Tour

In this area of central Kentucky, the horse is king. Visitors who wind their way through these country lanes will pass the black picket fences of fabled farms such as Claiborne and Calumet, home to many Kentucky Derby winners. Tours of some horse farms are offered and for those who would like to study thoroughbreds a little more up close and personal, the Kentucky Horse Park sits in nearby Georgetown. Keeneland is a short drive from many of these wineries, and the expansive grounds feel as if you're strolling through another era in Kentucky's history. The Red Mile Racetrack is also available to Lexington visitors, and a rare round barn is preserved on the property, so as not to lose this once common part of our culture.

Even the most simple can be revealed beautifully in the softly rolling landscape as hay bales and dandelions compete for attention in the bright sunshine. As visitors drive past this historic and breathtaking scenery, they'll need to keep a sharp eye out for winery signs. This is where an up to date GPS is essential, as many of the wineries only have the sign signifying people have arrived at their destination. Also, a few of them need to be followed by the directions on the winery's webpage, as a GPS will take you on a wild goose chase-been there, done that. Happy Wine*ing* and don't pass up the chance to taste the Black Barrel Reserve at Wildside Winery—it's phenomenal!

TALON TASTING ROOM • *Shelbyville*

RISING SONS HOME FARM WINERY • *Lawrenceburg*

LOVERS LEAP WINERY • *Lawrenceburg*

PRODIGY VINEYARDS & WINERY • *Frankfort*

EQUUS RUN VINEYARDS • *Midway*

BLACK BARN WINERY • *Lexington*

CASTLE HILL FARM WINERY • *Versailles*

WILDSIDE WINERY & VINEYARD • *Versailles*

TALON WINERY TASTING ROOM
SHELBYVILLE

Some of the wineries have tastings and sales at both the winery and a separate tasting room or satellite store, Talon Winery and Vineyard is one of them. Talon tasting room's wood and stone, cabin-themed structure, is a beautiful place to see. Inside, visitors will find a charming gift shop, as well as award winning wines.

The unique barrel shaped ceiling in the tasting room is a work of genius architecture. Outside, a back deck overlooks a sparkling pond. Visitors may glimpse ducks and many bird species as they sip wine and take in the rural surroundings. Music is offered every Saturday night. This is one of Kentucky's treasures that is simply not to be missed.

For Talon Winery's full story and recipe, please see page 164.

Talon Winery Tasting Room, Shelbyville
400 Gordon Lane
Shelbyville, KY 40065
502-633-6969

Events: Yes, indoor seating for up to 35, outdoor seating for up to 50.

Wines

⟿ **Southern Rosé**

This dry rosé is a delight with just a hint of sweet strawberry on the finish. A great wine to pair with pasta salad, grilled fish, or a plate of goat and feta cheeses.

⟿ **Toasted Bourbon Barrel Moonshine Jug**

Bourbon barrel-aged Chambourcin with dark fruit, caramel, and molasses aromas. The smooth finish of this wine pairs well with savory foods such as a spinach quiche or curried beef.

⟿ **Coyote Red**

A good soft red to serve with cheeses such as medium cheddar or Havarti, this wine also pairs well with fruit plates or is wonderful to drink by itself.

⟿ **Syrah**

An intense wine with a smooth finish, this slightly spicy wine has very fine tannins. This is the wine to serve at a backyard barbecue, as smoky meats pair well with the Syrah.

RISING SONS HOME FARM WINERY
LAWRENCEBURG

"Relaxing, Scenic, and Fun!"
– Francine Sloan –

In 1999, Francine and Joe Sloan established a vineyard on their property close to I-64 in Anderson County. The unusual name comes from the Sloans' three sons, all who don't especially greet the sun with enthusiasm. "We thought it important to have some agricultural work for growing teenagers, i.e.

pulling weeds," Francine said. The Sloans kept their full-time jobs while using some of the grapes for their own wine, jelly, and other products. They sold the remainder to others, but they always had their eye on opening a winery when their lives slowed down a bit. "Making wine has been a part of Joe's family since before his grandfather immigrated to America from Italy in the 1920s," Francine explained. "It was a natural progression from making our own wine to introducing our product to the public." Francine, a former school teacher, has a natural way with people and welcomes visitors as if they were lifelong friends. She makes most of the unique gifts in her gift shop and many of them are other items given a new life by Francine's imagination.

Francine and Joe Sloan have created a wonderful gathering place for family, friends, and friends they just haven't made yet. Photographs of grandparents, parents, and children of Joe and Francine hang on the walls of the Tuscan inspired tasting room. Keeping family close is very important to them.

Rising Sons Home Farm Winery
975 Frankfort Road
Lawrenceburg, KY 40342
502-600-0224

Events: Yes, outdoor seating for up to 250.

The art of winemaking is a custom in the Sloan family, and they use this opportunity to pay tribute to family members. The wine, Cataldo's Salute, is a nod to Joe's Grandfather, Cataldo Marinaro, who taught him how to give the toast, or salute, as it's known in Italy. Joe carries on the tradition within his family and, if prompted, may share this with visitors.

The Sloans introduce people to each wine as if it is a person with whom they will want to become acquainted, and they will. It is evident from the first sip that each wine is crafted with the person who will drink it in mind. Smooth and never overpowering on any one element, these are some of the most well-balanced wines in Kentucky. They're all so good it's very hard to pick a favorite. But it's fun to try.

121

Wines

❧ Norton

Dry and tart, this wine pairs well with roast beef or lamb. It will add interesting accompaniment to mild cheeses and tangy fruits such as cherries or plums.

❧ Chambourcin

Dry Earthy with a hint of spice. Works well with lamb, goose, or duck.

❧ Cataldo's Salute

Dry and made from the Norton grape, this wine is mellow and oaked with a hint of tobacco. Serve with spicy meats such as salami and pepperoni.

❧ Cabernet Franc

Dry with a slight peppery note. Pairs well with ham or roast pork.

❧ Baco Noir

Semi-sweet and slightly oaked. This wine will go nicely with oriental or Indian dishes.

❧ Bellissimo Bianco

A sweet white with a light, fruity taste. This wine is great to sip by itself or with a strawberry or other fruit dessert.

❧ Blackberry

A sweet dessert wine. Although many Kentucky wineries have a blackberry wine, I've yet to sample two that taste alike. This one goes well with medium cheeses.

❧ Vignoles

A semi-sweet wine with a crisp flavor. This one will make a delicious white sangria or try the wine bread recipe.

Recipe

Vignoles Wine Bread

2 cups Rising Sons Vignoles, heated 30-45
   seconds in the microwave
4 cups self-rising flour
4 Tbls. sugar
1 stick of butter

This is a great, easy, and tasty recipe. Although these directions are for a 9x13 pan, it can easily be halved for an 8x8 pan. Set oven to 350°. Place stick of butter into greased pan and set in oven. Stir flour and sugar together with wooden spoon. Heat wine 30-45 seconds, until slightly warm to the touch and add to flour mixture. Scrape sides to moisten all dry ingredients, then stir in the center of dough with wooden spoon until well blended and batter pulls away from sides to form a ball. Batter will be very thick and sticky. Scrape batter into melted butter in warmed 9x13 pan, patting gently to even. Bake for 45-60 minutes. Let sit for five minutes, slice and enjoy. Bread is the necessity of life. Wine is a close second.

*Don't make this
too early in the day,
or there won't be
any left for dinner.
It's addicting!*

123

LOVERS LEAP WINERY
LAWRENCEBURG

"All I want to do is make people smile."
– Brian Sivinski –

We both went to the polls and voted early the morning we went to Lovers Leap, then headed out to get some wineries under our belts, so to speak. We should have called first. In theory, we both know that alcohol isn't served until after the polls close on Election Day, but in our excitement, it just slipped our minds that we wouldn't be able to taste the wines. It turned out to be a bonus day. There were no crowds so it was easy to visit and ask questions, and returned another day to sip wine at this lovely winery. Brian Sivinski sat with us and discussed his thoughts and philosophies about wine and owning a winery.

"It's a lot of work, but a fun business to be in. No wine is ever the same. It's a living thing." The Sivinski's saw the potential of owning a winery and purchased Lovers Leap in 2010. When asked about the name of the winery, Sivinski imparted a local story.

Supposedly, two lovers whose parents forbade them to marry or even see each other again, joined hands and, literally, "took the plunge" on a ridge that overlooks the winery, a la Romeo and Juliette. Legends like this dot the Kentucky landscape and are part of the culture when traversing the state.

While Sivinski is pleased with his wines, his eye is on the wine industry 15-20 years in the future. "Quality control is a big deal in wine. Pruned properly, grapes give a smaller yield, but a better quality wine." Looking out over the expansive vine-

Lovers Leap Winery
1180 Lanes Mill Road
Lawrenceburg, KY 40342
502-839-1299

Events: Yes, indoor seating for up to
  90, outdoor seating for up to 200.

Above: Photo credit, April Cole.

yards, one can imagine the task of pruning on this large farm. Sivinski is also working on establishing an American Viticulture Society area in Kentucky. This would make Kentucky a designated wine grape-growing region and would bring both recognition and jobs as the call for Kentucky grown grapes would reach across the United States and beyond.

The wines at Lovers Leap are carefully chosen by health of the grape, customer taste, and quality control. Bring some favorite foods and taste Brian Sivinski's passion. He hopes to bring a smile to your face soon.

125

Wines

❧ Cynthiana

This gold medal winner is made from Norton grapes grown completely on the estate. Fig, currant, and spice combine in an oaked envelope of flavor. Rich, red, and completely Bluegrass. Pair with your favorite cut of steak or with a delicious corned beef.

❧ Merlot

French and American aged with bountiful fruit overtones of wildberry, black cherry and cinnamon.

❧ Photo Finish

A beautiful rosé made with Cayuga and Chambourcin grapes grown on the property. A great wine to pair with burgers and ribs hot off the grill.

❧ Win-Place Show White

This is blend of estate grapes which provides a delicate floral nose, light body and a smooth finish.

❧ Riesling

Slightly sweet and sultry, this is a good wine to serve with a group because it satisfies many palates. Pair with appetizers containing chicken, cream cheese, and mild cheeses.

❧ Photo Finnish Sweet Red

Medium bodied wine with hints of redcurrant, rasberry going 'nose to nose' with a sweet cinnamon closing in for a fruitful photo finish.

Recipe

Fish with White Reisling Sauce

1 cup Lovers Leap Riesling
Any thick white fish fillets, we used Swai
Garlic salt
Pepper
1 Tbls. corn starch, mixed until smooth in 1 Tbls. cool water
Lemon pepper, for garnish
Paprika, for garnish
Parsley, for garnish

Preheat oven to 350°. Place fish filets side by side in a baking dish lightly coated with cooking spray. Season with garlic salt and pepper. Pour the Riesling over the fish. Bake for 20 to 30 minutes. The fish is done when it flakes easily with a fork. Remove the fish from the baking dish and wrap it in foil to keep warm. Pour the wine mixture from the baking dish into a saucepan, and bring to a boil. Add cornstarch and water mixture, stirring until slightly thickened, 1-2 minutes. Pour immediately over the fish. Garnish with lemon pepper, parsley and paprika.

PRODIGY VINEYARDS & WINERY
FRANKFORT

"Deo Volente" (God being willing)
– *Chad Peach* –

Heading down 64 East toward Frankfort, visitors will find this quaint tasting room and gift shop. The tasting room is not open on Sundays, however, vineyard tours are only offered on Sundays, and we highly suggest you visit the lovely grounds where Prodigy's wines are produced.

The outside of this winery is unassuming and gives no clues to the inside atmosphere. A cute boutique offers fun finds such as Derby style hats, custom made metal wine caddies and napkin holders, and pretty hand-painted wine glasses. It has the feel of a wine bar with large televisions showing sports while energetic employees offer wine tastings.

Soon after marrying, owner Chad Peach and his wife Lenee purchased 63 acres in Woodford County. Their intention was always to move to the farm and grow something to help keep the agricultural spirit of the area alive and well. Wineries were just beginning to pop up around the state and in 1998, they planted their vineyard. Their plan was to grow grapes to sell to wineries, but they became curious. So many of the people they sold grapes to were raving about the quality of their vines that Chad tried his hand at making wine, "for a hobby."

Prodigy Vineyards & Winery
3445 Versailles Road, Suite A
Frankfort, KY 40601
502-352-9400

Events: Yes, at the vineyard, call for arrangements.

It wasn't long before he and his wife decided to open a winery of their own. Chad hired Jim Wight from Wight-Meyer Winery to help him make and produce wine worth selling. In 2010, Prodigy Vineyards and Winery opened their doors. While researching the land where the farm sat, the Peach's discovered many race horses were previously raised on their piece of property and decided on the name Prodigy.

As part of their winemaking philosophy, the Peaches like to keep the production as close to the winery as possible, even hiring a local artist to paint the thoroughbred horse on their signature label. They're pretty proud of it, as it won the "Best Label" competition at the Ag Genius Wine Label Competition, sponsored by the Kentucky Vineyard Society. In addition, their wines have won several medals at Kentucky festivals and the Kentucky State Fair.

On Sundays during the summer, the vineyard is open for a self-guided tour. While doing this, visitors may observe a myriad of animals including dogs, cats, horses, peacocks, chickens, and geese. The Peaches don't own the geese. There are two ponds on the farm and the geese adopted them. The winery hosts musicians on Fridays throughout the year, drawing a lot of people to their wonderful place.

Bring some companions, some tasty food, and get to know the good people at Prodigy sometime soon.

Wines

❧ Lemberger (Estate)

Barrel aged for nine months this wine contains flavors of black currant and cherry. Pair with creamy or cheesy pasta dishes, smoked meats or dark chocolate desserts.

❧ Diamond

A sweet varietal with tropical fruit aromas and flavors of pineapple, peach, and apricot. The perfect wine to pair with a spring chicken salad brunch or an appetizer party.

❧ St. Vincent-Cayuga Rosé

A blend of three wines, this unique offering suggests a hint of blackberry. This pairs well with non-smoked gouda, muenster and Havarti cheeses.

❧ Raspberry

Sweet fruit lingers on the palate with this gold medal winning wine. Pair with both sweet and tart tastes such as a lemon tart, coconut and chocolate cream pies, and cream puffs.

Recipe

Corny Chicken

"What is the definition of a good wine?
It should start and end with a smile."
– Willian Sokolin –

½ cup Prodigy Vineyards and Winery's St. Vincent
    Cayuga Wine, divided
1 package Jiffy Corn Muffin mix
2-3 pound chicken, shredded (I buy a rotisserie chicken
    already cooked and shred it.)
¼ cup olive oil
1 cup heavy cream
2 cloves garlic, crushed
1 small, sweet onion
1 small poblano pepper, chopped fine with seeds and
    stems removed (so it's flavorful, but not spicy-hot)
8-oz. container fresh sliced mushrooms
Freshly ground salt and pepper to taste

Mix corn muffin mix according to package directions. Add water a few tablespoons at a time until the batter is thin enough to pour easily. Add poblano pepper to corn muffin mixture. Heat olive oil in skillet and pour about 4 tablespoons of corn muffin mixture for each corn cake into hot oil; this should make 8 corn cakes. Remove to oven and keep warm. Pour all but two tablespoons of oil out of pan; add onion, mushrooms and garlic, cooking until soft. Remove vegetable mixture to plate. Add ¼ cup of wine to pan and stir the good bits from the bottom and sides of pan. When the wine has almost evaporated, stir in cream and remaining wine, then bring to a simmer, but do not boil. Season with salt and pepper. Return vegetables to pan until hot. To assemble, put two corn cakes down on plate and top with ½ cup of shredded chicken. Pour cup of gravy mixture over the top and serve immediately. Makes 4 hearty servings.

EQUUS RUN VINEYARDS
MIDWAY

"Visit the vineyard, taste the experience."
– Cynthia Bohn –

Kentucky Horse Park and Keeneland are two big attractions in this region of Kentucky. While driving along I-64, elegant horse farms can be seen, many offering tours. In addition, bourbon distilleries and trails abound here. This piece of Kentucky may seem to be all about horses and bourbon, but don't ignore the wineries of the area, they're more than worthy of a look.

The stories of Equus Run and its owner Cynthia Bohn, who also carries the titles of wine-maker, vineyard manager, and captain of the team, are respectively unique and prime examples

of the Kentucky spirit. Ms. Bohn believes the history of place is important and should be preserved. Morris Mill was a water powered grist mill that once sat on the property. The process is now mechanized but mills were a common sight in any community at the turn of the 20th century for their necessary use of crushing grains into flours and corn into meal. The remains of this once essential part of life can be seen along the creek just beyond the amphitheater at Equus Run Winery. A distillery also sat on the same land until about 1958.

Bohn worked for IBM for 30 years and started the winery in 1998 as a retirement investment, while still working full-time. Many Kentuckians believe farmland should yield something productive, and Bohn is no exception. "I grew up on a Kentucky tobacco farm and was happy to leave the corporate world and return solely to farming."

During her first year of wine production in 1999, Equus Run sold 450 cases. This year, they expect to produce around 9,000 cases. Wine is not the only thing Ms. Bohn has expanded since opening. "We've added the event barn, stage for the amphitheater, and many varieties of wines since we first opened," she explained. "We're always looking for the next thing we want to do." When looking for the next thing to do, Bohn keeps aesthetics in mind. Many buildings are works of art and creativity can be seen across the landscape including a bottle tree and artistic horses.

Equus Run Vineyards
1280 Moores Mill Road
Midway, KY 40347
859-846-9463

Events: Yes, indoor seating for up to
   250, outdoor seating for up to 1,200.

We noticed flowers planted at many of the vineyards and Equus Run had a small rosebush planted at the head of many rows of grapes. Although lovely, they're not just for looks. They attract bees which then pollinate the grapes as well as the flowers.

Bohn has never been one to shy away from work and visitors will find her in-volved in all job descriptions needed to run a winery, even sitting atop a back-hoe when necessary. Her hard work has brought national recognition to Bohn's vision. The winery has become a showplace which CNN travel de-scribed as a "Must see Hidden Treasure of the U.S." The picturesque setting of-fers a creek for fishing and there is a self-guided walking tour with descrip-tions of all the buildings on the prop-erty. Along with the wine, the merlot chocolate sauce is not to be missed. It's supposedly for ice cream, but a spoon-ful alone works well too. Plan a visit to Equus Run and "Taste the experience."

Wines

❧ Chardonnay

This medium-bodied wine has a crisp apple and lemongrass finish. Serve with a brie or camembert cheese dolloped with apple or pear butter and a sprinkle of golden raisins and sunflower seeds.

❧ Vidal Blanc

This wine bursts with pear, citrus, and granny smith apple flavors. A flowery aroma accompanies this light and lively wine. Drink alone or with grilled chicken for a little zing!

❧ Derby Bluegrass Blush

A varietal blend of aromatic whites and a touch of red with strawberry and raspberry scents with a very sweet finish. It's wonderful with lemon pound cake.

❧ Late Harvest Merlot

Rich raisin aromas with notes of clove. It's similar to a port.

❧ Passionate Kiss

Chocolate infused Cabernet Sauvignon. This wine has deep chocolate aromas with a cherry and strawberry finish. Great to sip alone or with a favorite fruit tart.

Recipe

Creamy Artichoke Rice in White Wine Sauce

½ cup Equus Run Vidal Blanc

16 ounces Long Grain Wild Rice, cooked and drained, reserve
  water

14-oz. can artichoke hearts, drained

8-oz. container cream cheese and chive

Dash of cayenne pepper

2 Tbls. fresh rosemary

Salt and pepper to taste

3 Tbls. butter

1 cup mozzarella cheese

1 Tbl. sunflower seeds

Parmesan Cheese

Melt butter in large saucepan over medium-high heat. Add mozzarella cheese, spices, and wine. Once cheese has melted and is hot, add rice and artichokes. If sauce is too thick, thin with reserved rice water, a few tablespoons at a time. Sprinkle with sunflower seeds and parmesan cheese. Serve warm as a side dish or add chicken and spinach and serve as a main dish.

Wine is the answer.
What was the question?

BLACK BARN® WINERY
LEXINGTON

"We craft one wine."
– Collin Boyd –

Visitors pass what seem like endless miles of black fencing on their way to Black Barn® Winery, but it's really only a few miles off the beaten track. Don't let the idea of making an appointment to taste this wine deter you, it's definitely worth the opportunity to visit owner, Collin Boyd. The trip to this memorable winery will stay with you for days to come. Mr. Boyd has such interesting stories and personable manners, that you could easily spend the day with him, and the night if you imbibe just a bit too much. I'm sure he'd offer a hay bale and a horse blanket to keep warm.

One of his interesting stories is about how the vintner of a small, family-owned winery in central Kentucky accidently became a winemaker while getting acquainted with royalty. "My uncle had a friend who lived in France and owned a zoo in Paris." Collin Boyd explained with a smile. His plan was to go to Paris the summer before starting college, work at the zoo and travel France. One problem though was his uncle could not remember the name of the zoo. Mr. Boyd went to France and found what he thought would be the right place, a Paris zoo owned by an American, but it wasn't. He called his uncle. Through friends, Boyd's uncle learned the zoo owner had become ill, quickly sold the zoo, and returned to America. Boyds's uncle had another friend living in France, and he owned a winery. "It wasn't much of a decision really," Boyd revealed while

holding his hands like scales. "Spend the summer shoveling animal manure or spend a summer growing grapes and learning about wine making." That family friend just happened to be royalty, Count Raban Adelmann originally from Germany. He

Black Barn® Winery
4200 Newtown Pike
Lexington, KY 40511
859-552-2525

Hours: By appointment only.

Events: No

even has his own Wikipedia page. (translate.google.com/translate?hl=en&sl=de&u=http://de.wikipedia.org/wiki/ Raban_Graf_Adelmann&prev=search)

The winery, Chateaux Canon La Gaffeliere, is still owned by royalty, Count Neipperg, and is situated in the fabled Saint Émilion wine region of France. The history of this region dates back to ancient times with vineyards planted as early as the 2nd century. (en.wikipedia.org/wiki/ Ch%C3%A2teau_Canon-la-Gaffeli%C3%A8re)

In many European locales, they make one type of wine and that area becomes associated with the name of the wine, Bordeaux, for example. This is the winemaking that Mr. Boyd emulates-he is working at perfecting one wine. It is a bold, dry wine, reminiscent of Cabernet Sauvignon. Instead of a name, his labels have a Roman numeral. These represent the number of times he has crafted his wine; IX was released in August 2012, X in 2013, and so on. Even though he only produces one type of wine, the taste still differs year to year. There are so many factors that affect the wine making process that, while similar, each year's vintage has its own unique flavor.

Mr. Boyd prefers grapes grown in California's Sierra Nevada Mountains because the region so closely resembles the climate of Southern France. When the grapes are close to being shipped, Boyd will put the word out and relatives, neighbors, and volunteers arrive to help with the harvest. Call and make an appointment to listen to stories and have a sip. It will only take one taste to convince visitors that Collin Boyd knows what he is doing.

Wine

❧ Black Barn Wine

Dry and traditional, this is the wine to serve
with everything from an elegant crown roast
or leg of lamb, to grilled burgers and steaks.

Recipe

Black Barn Burgoo

2 cups Black Barn Wine

3 lbs. cubed mixed meats (beef, lamb, pork, and/or chicken)

50 oz. frozen stew or soup vegetables, including sliced okra

3 cups chicken broth

2 cups beef broth

¼ cup Worcestershire sauce

2 small cans tomato paste

14-oz. can Italian style diced tomatoes

Salt and pepper to taste

Layer ingredients in slow cooker, starting with meat followed by the frozen vegetables, liquids and ending with tomatoes. Cover and cook on low heat for 8 to 9 hours until meats and vegetables are tender. Add salt and pepper to taste. This soup is great on a cold Kentucky night, especially if served with hot sauce and a glass of Black Barn Wine. Hint: Put all ingredients into the crockpot the night before and set in refrigerator. Turn on in the morning and let simmer all day.

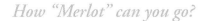

How "Merlot" can you go?

CASTLE HILL FARM WINERY
VERSAILLES

"A Little Piece of Heaven."
– Kim Addams –

As visitors to this region of Kentucky drive past curving black fences, they will understand why this part of the book is called "The Thoroughbred Tour." People may not realize the area is also known for the surrounding towns and architecture. Pisgah is one of the largest historic districts in the U.S. Midway, Nonesuch, and Versailles are filled with antebellum homes and antique shops, while some of the bed and breakfasts in this area date to the 1700s. The drive to this winery is filled with history, but there's mystery too.

Rumors abound about the Martin Castle/CastlePost structure that sits on the property adjacent to Castle Hill Winery. One story goes that Rex Martin made money as a coal baron in the 1970s and started building the castle to honor a promise made to his wife while touring European castles on their honeymoon. It is rumored that when they divorced, Martin was heartbroken and refused to finish it. Questionable ownership by famous people, a mysterious fire, and an eventual opening to the public, have made the castle a little more accessible. Despite public access, it has kept many of its secrets, remaining an enigma for those who drive past the closed gates of the castle on the hill.

Upon entering Castle Hill Winery, the visitor knows one thing for certain; owner Kim Addams is nostalgic. A 1950s' era blue bicycle hanging on the wall, used to belong to her mother. When her father found out that Kim's mother never owned a bicycle, he salvaged 3 bikes and built one for her. Addams and

Castle Hill Farm Winery
3650 Lexington Road
Versailles, KY 40383
859-576-0010

Events: Yes, indoor seating for up to
200, outdoor seating for up to 300.

her mother spent many happy hours riding together. The bike had such significance that she could not part with it, so now it has a place of prominence and catches visitors' eyes as they walk through the entrance.

Kim's love of reusing all that is old is plainly seen throughout the winery. Antique plates and glassware cover the tops of many vintage tables sitting around the gift shop. It's beautiful and sentimental, giving the visitor the feel of an old time country store. The bricks used for the tasting room walls came from an old barn and the doors leading from the gift shop to the tasting room used to be the entrance to the ball room in a grand Lexington mansion. Even the property the winery sits on is historic, once belonging to Kentucky's 58th governor, Brereton Jones.

On a hot summer day, the expansive front porch and side patio are perfect places to bring some friends, favorite foods, and sip a glass of wine, or three.

141

Wines

ꝥ Shiraz

Nice and spicy with hints of blackberry and plum. This wine goes well with hearty beef dishes, plus Indian and Latin cuisines, and other spicy foods.

ꝥ Castle Hill Red

Mellow, yet bold with plum flavors. Works well with marinara sauces, beef, and Mediterranean seasonings.

ꝥ Castle Hill White

Sipping this semi-sweet wine is like biting into crisp fruit. It would pair well with a cheese tray or chicken dishes with Mornay or Alfredo sauce.

ꝥ Camelot

A sweet white wine with a hint of vanilla and tropical fruits. Wonderful chilled and paired with a plate of light flavored cheeses.

5

Recipe

Can't be Crabby Salad
with Warm Wine Dressing

½ cup Castle Hill White

6 Tbls. Butter, divided

2 scallions, sliced

Juice from one lime

4 Tbls. heavy cream

1 small container of crumbled feta cheese with
  Mediterranean spices

4 pieces of bacon, fried and crumbled

Salt and pepper to taste

One bunch romaine lettuce, washed and divided among
  4 plates

16 ounces imitation crab meat

Melt 2 Tbls. butter in heavy skillet add scallions and crab meat until seared. Remove crab meat to plate and keep warm. Add lime juice and wine to skillet and cook until reduced by 90%. Add heavy cream. Cut remaining butter into slices and whisk until butter is incorporated. Salt and pepper to taste. Sprinkle bacon and feta cheese evenly over lettuce. Divide crab evenly over lettuce and pour hot dressing over each. Serve immediately.

WILDSIDE WINERY & VINEYARD
VERSAILLES

"Explore your Wildside!"
– Neil and Rachel Vasilakes –

Just three and a half miles off the Bluegrass Parkway, among the fertile farmlands of central Kentucky, sits a winery that is close to everything, but worlds away at the same time—Wildside Winery and Vineyards.

Owners Rachel and Neil Vasilakes began their winery path when they bought this farm in 1997. "Right away we planted

grapes, apples, peaches, and berries," said Rachel. They wanted to develop wines using as many organic methods as possible. After the first planting, the Vasilakes added more fruit each year until they now have eleven acres of grapes, two acres of fruit trees, and an acre and a half of berries.

"I started as a grape grower and amateur wine maker," Neil said. "I then headed up a private wine making club. After several hundred batches of wine, we got our commercial winery license in 2004."

With Neil as the vineyard manager and wine maker, their selection has expanded until there are over 25 varieties. He likes to try making both traditional and unusual types of wine. Some of the varieties visitors will find are Black Barrel Reserve, a dry Cabernet aged in Woodford Reserve Bourbon Barrels, a semi-sweet Paw-Paw, a sweet Black Walnut, and Peach Mead. Neil's sense of humor shines through in a wine named Dangerous. "I named it that because it's so good you're in danger of drinking the whole bottle."

Visitors to this winery will appreciate the outlook of these winemaker/owners. "Our

Wildside Winery & Vineyard
5500 Troy Pike
Versailles, KY 40383
859-879-3982

Events: Yes, indoor seating for up to
175, outdoor seating for up to 200.

mission is quality wines at a reasonable price," Neil explained. Another objective is to have "serious, undiluted, pure fruit wines all made from 100% fresh Kentucky fruit." Wildside qualifies as a Kentucky Proud home farm winery.

Wildside Winery offers many picturesque spots to enjoy a Kentucky day. The pavilion is big enough for a crowd. Popular music events are held regularly during the warmer months.

Mr. and Mrs. Vasilakes are portraits of hard working Kentuckians. They both have other full-time jobs yet work hard at maintaining the vineyards, the winery, and their friendships. One employee explained, "I've worked here for four or five years and have enjoyed every minute of it. They're not employers, they're friends. They're just wonderful!"

We believe anyone who enters their doors, tastes their wines, and is a guest to their wonderful hospitality, will agree with this assessment.

Wines

Viognier

A dry white with fruity characteristics, serve this wine with Moroccan, Indian, and Caribbean spiced dishes, or a tray of soft cheeses.

Duet

This dry red blend of Foch and DeChaunac grapes, is estate grown. Pair with a rack of lamb, red pasta sauces, and spicy dishes.

BBR—Black Barrel Reserve

A Cabernet Sauvignon and Chambourcin blend, this wine is aged in Woodford Reserve Bourbon Barrels for over a year. A best seller to pair with scallops and fried green tomatoes.

Frontenac Gris

This grape makes a delicious, semi-sweet wine perfect for sipping chilled on a hot summer evening.

Black Walnut

Sweet and strong, this wine would be excellent added to fudge or poured over ice cream.

Recipe

Becky's Almost Famous Baked Beans

½ cup Wildside Winery & Vineyard Dangerous Wine
5 slices smoked bacon
1 cup light brown sugar
3 tsp. yellow mustard
Five 14.5-oz. cans of pork and beans, drained with pork fat removed

In large microwave-safe dish, fry bacon until crisp, drain on paper towels. Add wine, light brown sugar and mustard. Stir and cook 3 minutes on high. Stir in drained beans and cook for 20 minutes on high. Remove carefully, stir and allow to cool slightly. Crumble bacon on top. These stay warm for a long time, so they're perfect to take to a party.

Whenever going to a family get-together, someone always says; "Bring your beans, they're like dessert," or "Aunt Becky's beans are the best, and I don't even like baked beans." One of the best parts of this recipe is it's all cooked in the microwave. Here's the secret.

LEXINGTON AREA

LEXINGTON • WINCHESTER • NICHOLASVILLE • RICHMOND • DANVILLE

The Stone Fence Tour

Stone fences line many of the streets visitors will wind around to find the wineries in this area of the state. Sometimes called rock fences, there are more of these in central Kentucky than anywhere else in the United States (lizettefitzpatrick.com). These originated in the Bluegrass when the Irish Potato Famine of the 1840s, forced many people to immigrate to the U.S. and some settled in central Kentucky. Irish stonemasons built many of the fences still seen today throughout this region. Their craft of stacking Kentucky limestone, often found in fields and riverbeds, with no mortars or cement, is quickly becoming a lost art. Luckily, there are a few places which offer classes for those wishing to learn more about building this type of structure and a movement has started to restore and preserve the existing fences.

Many of the wineries in this area lie within, or very close, to Lexington city limits. One bonus to the abundances of this region is an array of free activities. At the Art Museum at the University of Kentucky, traveling exhibits allow for a different experience upon each visit. Take a plant walk across Kentucky at tha Arboretum on Alumni Drive or stroll through the gardens of Lexington Cemetery.

Raven Run Nature Sanctuary offers visitors the chance to commune with nature and Legacy Trail is a nine mile bike trail with safe places for children to learn to ride. Jacobson and Shillito Parks' creative playground structures are second to none. Sweet treats and free samples can be found at Old Kentucky Chocolates and the Lexington Public Library displays the world's largest ceiling clock. Whether the goal is to be on the move, sit back and relax, or both, offerings abound in this area of Kentucky.

CHRISMAN MILL WINERY STORE • *Lexington*

HARKNESS EDWARDS VINEYARDS • *Winchester*

GRIMES MILL WINERY • *Lexington*

JAMES FARRIS WINERY & BISTRO • *Lexington*

TALON WINERY • *Lexington*

CHRISMAN MILL VINEYARDS & WINERY • *Nicholasville*

ACRES OF LAND WINERY • *Richmond*

FIRST VINEYARD WINERY • *Nicholasville*

CHATEAU DU VIEUX CORBEAU WINERY • *Danville*

CHRISMAN MILL WINERY STORE
LEXINGTON

This is the tasting room and gift shop for Chrisman Mill Vineyard and Winery in Nicholasville (see page 168 for Chrisman Mill's full story). In addition to visiting this tasting room with award winning wines and a quirky gift shop, we highly recommend visiting the lovely winery and vineyard in Nicholasville. Along with wine tastings, this Lexington location offers a house made tapas menu and craft beers. This beautiful space is in the very popular Hamburg Pavilion shopping area with many entertainment and dining selections. It would be easy to spend a whole day, or two, exploring all the offerings in this area of Lexington alone.

Chrisman Mill Winery Store
2300 Sir Barton Way #175
Lexington, KY 40509
859-264-9463

Events: No

151

HARKNESS EDWARDS VINEYARDS
WINCHESTER

"Each wine is an experience and the most romantic connection to our sacred land."
– The Edwards –

Unfortunately, on August 8, 2013, a lightning strike started a fire which destroyed the tasting barn and all of the wine in stock at this lovely winery. They do plan to rebuild, but it will take some time. Here is their story.

The land in this river valley area of Kentucky slopes gently and softly making it the perfect complement to grape vines. The soil variations in this portion of the state, with its abundance of limestone and clay, also lends itself to interesting combinations and textures for winemaking.

Owners of Harkness Edwards Vineyards, Harkness "Harky" Edwards and his wife Kathy, have a long history of farming. Harky and Kathy have called this area of Kentucky home for all of their lives, but never thought about growing grapes or owning a winery. When neighbors were selling their adjoining farm, they asked the Edwards if they would like to purchase it. After the purchase, they were discussing what agricultural crops would be best to grow. A friend suggested grapes and after some research about the feasibility of grape vines in Central Kentucky, their journey began.

On the day we visited, they were harvesting Cabernet Doré and we ate our fill of this wonderfully mysterious grape. The vine comes from mixing two deep red grapes, Cabernet Sauvignon and Norton's, to create a pale yellow fruit with a flavor all its own. Mr. Edwards allowed us to observe the process of

Harkness Edwards Vineyards
5199 Combs Ferry Road
Winchester, KY 40391
859-527-6635

Events: Not presently, but there are
plans to host events in the future.

moving grape to juice and gave us tastings of this magnificent addition to the vineyard.

He also took us to his weighing shed. The antique scale involved represents the Edward's appreciation for keeping things around based on their usefulness, not because a newer model may be available. Harky smiled broadly as he showed how to use the scale and explained that even though it's about one hundred years old, it's still accurate. Everywhere visitors look at this winery there is a perfect blend of old and new. The tasting room is a converted barn with a beautiful, renovated brick silo. The vineyards are laid out to draw the eye to the land and what may lie beyond

the boundary. Even the labels on the wine combine old world designs with new ideas to create something unique and exciting. Three of their labels recently won 1st, 2nd, and 3rd, respectively in a wine label contest, highlighting the care that goes into every aspect of Harkness Edward wines.

It has been a ten year journey for the Edwards from acquiring adjoining land to opening a winery for the public. Visit soon and taste the "romantic connection to our sacred land."

153

Wines

Vat 32 Viognier

Dry and crisp with fruity and floral aromas, this beautiful wine is the one to serve at a garden party with a Caribbean spiced salmon, crab salad, or honey glazed scallops.

Night Heron

A dry red Norton, this classic goes well with spicy or plain beef dishes. Add to beef or pork based soups to give them extra depth.

Taste the Sun

A semi-sweet wine bursting with fruitiness, this versatile wine goes well with white pasta sauces, chicken, turkey, and pork dishes.

Big Red

Named for the racehorse Man O' War, this sweet concord wine is full-flavored and wonderful to sip alone or with desserts.

Fling

Clean and smooth, this is the wine to share on a girls' movie night. Pair this wine with well-seasoned poultry dishes, or flavored popcorn. You can't have a movie night without popcorn.

Recipe

Deviled Quiche Lorraine Eggs

¼ cup Harkness Edwards Vat 32 Viognier
12 eggs
1 cup or 32 oz. real bacon bits
1 cup mayonnaise
1 cup shredded Italian style mozzarella cheese
1 tsp. paprika
1 Tbls. sugar
3 Tbls. Dijon mustard or one of your favorites
Paprika or parsley for garnish

Boil eggs ten minutes to hard boiled. Cool in ice water and peel immediately. Slice eggs length-wise, scoop out yellow, and place in bowl. Place twelve egg white halves in a deviled egg holder, reserve remaining egg whites. In medium bowl, mash all twelve yolks until smooth. Add bacon bits, mayonnaise, mozzarella cheese, paprika, sugar, wine, and mustard. Fill each egg in holder with two teaspoons of mixture, there will be leftover yolk mixture. Chop remaining egg whites and add to yolk mixture, mix well. If necessary, add a small amount of sugar, mustard, and/or wine, according to taste. This mixture can be used as a dip for crackers, chips, and vegetables.

My husband came from England many years ago. He introduces our family to a lot of interesting recipes and we add a bit of regional charachter. He loves Kentucky and being married to a "good ol' country gal." Along the way to gaining an American citizenship, he also acquried a slight southern twang.

GRIMES MILL WINERY
LEXINGTON

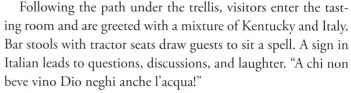

Winding your way through this part of Kentucky, her beautiful landscape is in its full glory. Nearby Jacobsen Park offers visitors picnic shelters, playgrounds, basketball courts, and a lake with paddleboat rentals. The historic site, Boone Station, is a place for quiet reflection of the sacrifices Daniel Boone and his family made for Kentucky. Not far from these two places is a spot Kentuckians, or anyone, will want to visit over and over again, Grimes Mill Winery.

Following the path under the trellis, visitors enter the tasting room and are greeted with a mixture of Kentucky and Italy. Bar stools with tractor seats draw guests to sit a spell. A sign in Italian leads to questions, discussions, and laughter. "A chi non beve vino Dio neghi anche l'acqua!"

"To who does not drink wine," co-owner Phillip DeSimone explained, "God denies also water!" He, his wife Lois, and their family lived in Italy for a year and really enjoyed the tradition of slowing down in the late afternoon with cheeses, breads, meats, friends, family and wine. They wanted to introduce this 2nd and 3rd generation family tradition to their adopted and beloved Kentucky home. With the idea of a winery in mind, the DeSimone's purchased a tobacco farm in 1997. The family decided to utilize the existing structures and kept the original name of the farm. Philip is very proud of both his Italian and Kentucky heritages.

"Grandfather Proto made wine in his Brooklyn (New York) basement, and shared it with family," he said. "Wine is a simple pleasure that brings family and friends together as they share a meal." His grandfather was pleased at the enthusiastic response to his homemade wine and called it Ca' Proto—from the house of Proto. The DeSimone's carry on this tradition, with

Grimes Mill Winery
6707 Grimes Mill Road
Lexington, KY 40515
859-543-9691

Events: Yes, indoor seating for up
to 30.

Ca' DeSimone on each label. Philip, also affectionately called Dr. D, shares his grandfather's passion and exuberance for wine.

"Here, taste this. You have to taste this one," he laughingly prodded as he took us on a personally guided tour. Grimes Mill offers visitors 2 classes, which include tours of the facility, titled Wine 101 and Wine 102, respectively. The DeSimone's love sharing their knowledge of family winemaking with the public, and they have an eye toward expanding that to other traditions in the future.

"We dream of expanding," Lois told us, "of putting in an outside wood burning oven for Italian breads and pizzas." Visitors are encouraged to bring their own food and let the DeSimone's match one or more of their wines with it. They believe wine should be enjoyed with food. Bring some favorite foods and visit this Kentucky jewel soon. We guarantee you'll be back.

Wines

☙ **Bianca**

Crisp with notes of citrus, this award-winning dry white, named after the DeSimone's dog, goes well with pasta in cream sauces and fish such as salmon.

☙ **Moscato**

This sweet white is a sip of cool on a hot summer day. Serve chilled with your favorite dessert or fruit salad.

☙ **Fantasia**

Semi-sweet and blush in color, this table wine is perfect to serve with mild cheeses.

☙ **Fioro Roso**

"Red Flower- this light and fruity blend pairs well with almost any food.

Recipe

*"Life is way too busy—sit back,
eat what you like,
drink what you love
and enjoy the game!"*

Take a "Cab" to the Game Chip Dip

½ cup Grimes Mill Cabernet Sauvignon

1 lb. lean ground beef

⅛ tsp. cayenne pepper

⅛ tsp. garlic powder

Pinch of salt

Pinch of pepper

8 oz. taco sauce, divided

16-oz. can refried beans, divided

2 avocados, peeled and seeded

Lime juice

1 cup chopped sweet onion

1 cup shredded Italian cheese

1 cup sour cream

15-oz. jar con queso

4 oz. sliced ripe olives

8 oz. cream cheese, softened

2 cups shredded Cheddar cheese

8 oz. shredded lettuce

1 medium tomato, chopped

Nacho chips

In a medium sauce pan, combine ground beef, cayenne pepper, garlic powder, salt, pepper and Cabernet Sauvignon. Simmer until beef is cooked thoroughly, stirring occasionally to break up beef. Once cooked, drain the beef; add ½ cup of the taco sauce and 4 oz. of the refried beans, warm slowly until mixed well. Set aside. Mash the avocados and add a splash of lime juice to preserve the green color. Cover and refrigerate until needed. Warm remaining refried beans until soft and spread over the bottom of 9x13 baking pan. Over the beans sprinkle onion, then Italian cheese. Spread sour cream over the Italian cheese and pour con queso over the sour cream, followed by the olives and cream cheese. Spread meat mixture over the cream cheese, followed by remaining taco sauce. Top with the prepared avocados and shredded cheddar cheese. Cover and refrigerate until ready to serve. When ready to serve, spread lettuce and tomatoes on top. Serve with the nacho or corn chips.

"Serving life, following passion"
– Ben O'Daniel –

No strangers to the wine industry, husband and wife owners, Jean and Ben O'Daniel, have been around grapes and wine from a young age. For a science experiment as a senior in high school, Jean planted several varieties of grapes on her parents' reclaimed Eastern Kentucky coal mine property, "just to see what would grow." Almost everything did.

Her parents, being natural Kentuckians, didn't want all those grapes to go to waste, so they started experimenting with wine making. They own Highland Winery, listed on page 266. Jean's parents' help and enthusiasm for another successful agricultural crop on their property fueled her interest in wine making and the feasibility of opening a winery in Kentucky. She enrolled in the bachelor of food science degree program at the University of Kentucky. Soon after completing the program, she took a serendipitous week-long bus tour of some Missouri wineries.

"There were two younger people on the bus, myself (Jean) and Ben." "Ben" was Ben O'Daniel, son of Eddie O'Daniel, owner of one of Kentucky's first wineries to open in the wave that began in the late 1990s, Springhill. Over the course of the week, they got to know each other as they sampled and discussed wines. Their singular vision of owning a winery was enhanced when they married, each bringing special talents to the project. They opened the winery in 2002 and their wines began winning national and international competitions. In 2012, Jean

Farris's 2007 Cabernet Sauvignon won double gold in the San Francisco Chronicle Wine Competition. Many wineries open a tasting room first, and then gradually expand to add more amenities. However, the O'Daniels added a restaurant from the beginning. Both coming from agricultural backgrounds, the O'Daniels are big believers in cutting down the amount of steps from farm to table. They scouted out heirloom toma-

Jean Farris Winery & Bistro
6825 Old Richmond Road
Lexington, KY 40515
859-263-9463 (WINE)

Events: Yes, business meeting style,
35-40 people.

toes and other vegetables and fruits to grow themselves for the restaurant. Chickens can be seen on the grounds, including one rooster who likes to stop and pose for pictures. The bistro menu changes often, highlighting what is in season. Wines are suggested throughout the meal, to complement the foods being served. Jean Farris Winery and Bistro is a complete food and wine experience for visitors.

One experience visitors seem to enjoy is the Movies in the Vineyard, held once a month, weather permitting. "We make specialty popcorns such as parmesan and truffle," Jean O'Daniel said. "People seem to really like it."

They seem to like a lot of things about Jean Farris Winery and Bistro, as there were many people sipping wine and eating on the outdoor patio, which affords lovely views of the garden, vineyards, barn, and a bank of trees in the distance. Visit this wonderful winery soon and soak in some of the best things Kentucky has to offer.

Wines

❧ Tempest
This dark, spicy blend has notes of tobacco and pairs well with lamb and beef dishes.

❧ Hell Hound Red
This wine is named for the winery dog, Hades. Robust with layers of cherry, vanilla, and spice. Serve it with red pasta sauces and spicy dishes.

❧ Marito White Kentucky
Clean and crisp, this fruity and floral blend is wonderful chilled for sipping or pair with a flaky fish, such as bass, sole, or trout.

❧ Riesling
Lightly sweet, this wine offers fruity overtones with a citrus finish. Sip and enjoy!

Recipe

Say "Cheese" Cake

11.2-oz. box no-bake cheese cake mix
8 oz. strawberry preserves
⅓ cup Marito White Kentucky Wine
Fresh strawberries for decoration and added strawberry
  flavor

Follow directions on the back of cheese cake mix box; set aside. In small sauce pan, slowly bring to boil strawberry preserves while stirring. Turn heat to low and add wine, stirring until desired thickness. You can add more wine or less, depending on your taste buds. Let cool completely, stir once more. Spread on top of cooled cheese cake. Top with straw-berries. Keep refrigerated.

"I gotta sweet tooth and strawberry youth."
– Jeffree Starr –

TALON WINERY & VINEYARD
LEXINGTON

"In Vino Varitas" (In wine, truth.)
– Alcaeus –

This is Talon's Lexington location, not to be confused with the Shelbyville tasting room. The history and beauty of this land with its smooth hills, proud trees, and lovely pond, drew owner Harriet Allen to purchase the former tobacco farm in 1997. The beautiful Kentucky landscape is complemented by a historic home on the property, which has been transformed into a tasting room and gift shop. Around 1790, the house was built by Kentucky's first governor, Isaac Shelby (1750-1826), as a wedding gift for his daughter.

The Homestead Tasting Room is lovingly restored and decorated with many period pieces. Since the view has changed little, visitors can take a glass of wine to the expansive side porch and daydream about what life was like in the late 1700s.

Current owner, Ms. Allen, planted the first vines in 2001, gradually increasing to the current five acres. Talon currently grows seven varieties of grapes: Chardonnel, Cayuga White, Traminette, Chambourcin, Cabernet Sauvignon, Cabernet Franc, and Vidal Blanc. Although these names may sound familiar to most wine drinkers, the combination of soil, growing season, blends, and fermenting practices make these wines taste different from other wineries' selections. This is the mark of a great winery, creating signature wines.

One way that this idea is clearly demonstrated is that each wine at Tal-

Talon Winery & Vineyard
7086 Tates Creek Road
Lexington, KY 40515
859-971-3214

Events: Yes, indoor and outdoor
seating for up to 200.

on has won medals at competitions all over the United States. This includes double gold for their Monarch in the 2012 Indianapolis International Wine Competition, gold for the Traminette at the Kentucky Fair Wine Competition, and gold for the Chambourcin at the 2011 Kentucky Derby Wine Competition. Sweet Evening Breeze is their bestseller, with Blackberry a close second.

One unique feature Talon offers is a self-guided tour around the property. This tour was created by winemaker Kerry Jolliffe, and highlights the entire winemaking process "from vine to bottle." The excursion takes visitors on a journey from the beginnings of the winery to its present day operations.

On the day we arrived, a wedding was taking place in the Moondance Gazebo, on a hill overlooking the pond. It was a beautiful scene for any who happened to stop by the winery that day. In addition to the gazebo, Talon offers the 4,500 square foot "Ca'Barn'et Barrel Barn" with adjoining patio for receptions and large parties. For smaller gatherings, up to 58, Talon has the Monarch Pavilion and Evening Breeze Gazebo. Whatever your need for a gathering or just a quiet afternoon with friends, Talon can accommodate.

Wines

❧ Moondance

A Pinot Gris with fantastic tropical flavors. It's great with salmon or tuna.

❧ Sweet Evening Breeze

A Riesling blend, semi-sweet and crisp with a delightful fruity finish. Bring this out with a lightly flavored cheese and meat tray.

❧ Monarch

Astonishing red color and a peppery flavor enliven this estate grown wine. Serve with your favorite grilled beef dishes, even hamburgers.

❧ Afterglow

A semi-sweet blend of Catawba and Concord, this wine is wonderfully reminiscent of Kentucky summer. Sip alone or with pork or beef dishes.

❧ Chocolate Strawberry

This sweet dessert wine is a treat all by itself.

Recipe

Pollo Barracho or Drunken Chicken

1 ½ cups Talon Winery's Moondance Wine, divided
2 to 3 lbs. shredded chicken (I buy 1 rotisserie chicken
   already cooked and shred it)
1 small can red ranchero sauce
3 cups shredded Monterey Jack Cheese
1 cup sour cream
Two 4-oz. cans diced mild green chilies, drained
8 large flour tortillas

In medium sauce pan, put chicken, ranchero sauce and ½ cup of wine and heat on medium high. In another sauce pan, place 1 cup of wine and heat for 4 minutes on medium-high. Reduce to medium-low and add cheese, stir until melted. Mix in sour cream and chilies, cook until heated through, but do not boil. Cook tortillas according to package directions. Divide chicken evenly between tortillas, roll up, and place on serving platter. Pour sauce over them and serve immediately. Be sure to only sip your wine while making this recipe. Too much of this dish paired with the wine will have you doing a Moon Dance!

CHRISMAN MILL VINEYARDS & WINERY
NICHOLASVILLE

"Unique. Personal. Fun."
– Chris and Denise Nelson –

To say that Chris and Denise Nelson are workaholics is an understatement. He has a successful career as a physician, while she has an architectural degree. Denise designed their winery to resemble a lovely and traditional Tuscan vineyard. In addition, she added a full chef's kitchen and took cooking courses to hone her skills with the tapas menu that is served both here at the vineyard and the tasting room location in Lexington.

Chris's vision of a grape vineyard was planted when he was given a trip to Napa Valley for a college graduation present. "I'd experimented with wine-making and became intrigued in the whole process of making good wine." He and Denise have traveled many areas of the globe to divine how different countries go about creating different types of wine. Their precision for winemaking is evident in every glass. The honey mead is smooth and sweet, like tasting pure honey. The Greeks called mead "The nectar of the gods." Chrisman Mill's blackberry wine rivals my mother's blackberry cobbler—nothing compares to my mother's blackberry cobbler.

Although the Nelson's are Kentucky transplants, their focus is creating strictly Kentucky wines. "I'm interested in showing what Kentucky can

do," Chris explained. "There are so many people saying you can't grow certain grapes or use certain drying processes in Kentucky." Chris believes with enough patience and determination, he can prove these theories wrong. Despite being resolute in building their winery to its greatest potential, the

Chrisman Mill Vineyards & Winery
2385 Chrisman Mill Road
Nicholasville, KY 40356
859-881-5007

Events: Yes, outdoor only. Future plans for an indoor venue.

Nelson's are extremely humble and gracious toward other wineries and the wine industry in Kentucky.

"Have you been to First Vineyard yet?" Chris inquired. "They have such an amazing story." Chrisman Mill even has a wine named "First Vineyard Reserve." Because of trying to recognize all the people who helped them get to where they are in the industry, it was hard to keep Chris focused on his story.

"We were the 5th winery opened in Kentucky after prohibition ended." Their first vintage was produced in 1999. "I'm pretty proud to be holding license number 11 in the state." Chrisman Mill focuses on growing Vidal Blanc, Cabernet Sauvignon, Chancellor, Cabernet Franc, and Norton grapes, to perfection. Only with the ripest grapes can one make the best wine." Come visit Chris and Denise at Chrisman Mill Vineyard and Winery, but be prepared for some new additions, these energetic people are always expanding.

Wines

Ensemble

Similar to a chardonnay, this oaked wine is a great accompaniment with garlicky shrimp and crab dishes or turkey.

Norton Reserve

Available in very limited quantities, this "big red" wine, meaning dryer with higher tannins, goes well with hearty beef dishes.

Sweet Jessamine Rose

Similar to a zinfandel with its pretty blush color, this wine pairs with barbecue, grilled meats, and red pasta sauces.

Sweet Riverbend Red

This gold medal winner is a 100% Kentucky blend. It's great for sipping alone, in a sangria, or making a frozen slushy for a hot summer evening.

Recipe

Sweet and Savory Phyllo Cups

½ cup Chrisman Mill Winery's Honey Mead
8 oz. cream cheese
4 Tbls. diced ham
2 tsp. chopped chives
¼ tsp. onion salt
¼ tsp. garlic powder
30 phyllo cups, found in freezer section

Mix all ingredients except phyllo cups. Carefully spoon ½ teaspoons of mixture into each cup. Bake at 350° for 10 minutes. Enjoy!

Phyllo dough, sometimes spelled fillo, is paper thin sheets of dough first created in Greece long ago. The dough is often seen in chocolate or cheesecake desserts, but it's yummy with savory dishes too.

ACRES OF LAND WINERY & VINEYARD
RICHMOND

A sip of Kentucky's wine history: NOW CLOSED

This part of Kentucky is all about the collegiate life. The University of Kentucky is spread across many acres around Lexington. Just down the road in Richmond, Eastern Kentucky University attracts many students from all over the United States each semester, and in Berea, Berea College sits as a model of hope for many students. Charging no tuition, Berea only accepts students with the highest academic standings. Each student is required to work on campus to help finance careers and create a sense of belonging to the college and community. Close to all of this sits Acres of Land Winery, a place with a business model many former tobacco farmers emulate.

In the 1940s, Lowell Land's parents, Russell and Willie Mae Land, purchased 400 acres to grow burley tobacco. This crop sustained and helped the farm to grow for many years, but when tobacco's future began to decline, Lowell and Katherine looked for ways to diversify the farm and keep their agricultural heritage. They planted their first one and a half acres of vines in 2000, with Reliance, Mars, Chambourcin and Vignoles grape varieties. The Lands have continued to improve and expand the winery until they have what they now call an "agritainment business." Acres of Land offerings also include a restaurant and brewery.

In 2005, a tobacco barn was renovated to hold a 150 seat restaurant, gift shop, and tasting room. "Once you try our de-

Acres of Land Winery & Vineyard
2285 Barnes Mill Road
Richmond, KY 40475
859-264-9463

Events: Yes, indoor seating for up to
300, outdoor seating for up to 300.

licious food, you will begin planning your next trip back. Our chef is second to none and the proof is in the entrée," said Lowell Land. The Land's work hard to promote a farm family atmosphere on their property. There's a garden from which they pick many of the vegetables used in the restaurant. In addition, they've set up picnic areas around the farmstead so visitors can enjoy the beauty of the countryside.

In 2007, the Land's added wagon tours to their offerings. Visitors can take them through the vineyard and around the property. The wagon tour will also drop visitors at various picnic spots to enjoy a lunch, packed by the restaurant, and a bottle of wine. There are tours of the winemaking facilities as well, for those who want a deeper look at this process.

The Lands are often asked for their secret to success. "It's simple really. We do things the Kentucky way with patience, craftsmanship, and pride." Visit Acres of Land soon and taste the success of hard work.

Wines

Phoenix

Just like the bird that arose from the ashes, this wine represents the restaurant's burning in 2009 and subsequent return. Serve with comfort foods like macaroni and cheese.

Kentucky Chambourcin

Bursting with fruity flavor and just a hint of spice, this wine would go well with hearty beef stews or chili.

Concord

Often cited as their best seller, this wine is full-flavored and best served chilled.

Maggie Adams Blush

Described as a "white concord" this sweet wine is great to sip or adds interesting texture to dishes such as lasagna with red or white sauce.

Willie Mae's Blackberry

Named for Lowell's mother, this is a sweet treat to savor.

Recipe

Inside Out Pizza

1 cup Acres of Land Marie's Merlot
1 can pasta sauce
1 pound ground chuck
1 canned thin crust pizza crust from the refrigerated section
Various pizza toppings, according to taste
1 ½ cups mozzarella cheese
Parmesan cheese
Italian seasoning

Mix sauce and wine, heat on medium high about 20 minutes or until thickened. Brown ground chuck and drain well, setting on paper towels to absorb as much oil as possible. Add to thickened sauce. Spread pizza crust on greased rectangular cookie sheet. Make 6 to 7 cuts about two inches wide and three inches long down each side of the crust. Pour 1 cup of sauce and meat down the center of crust. There will be extra sauce. Add toppings and mozzarella cheese. Fold strips to meet in the middle over cheese. Sprinkle with Parmesan cheese and Italian seasoning. Bake at 350° for 15 to 20 minutes or until browned. Serve with extra sauce for dipping.

FIRST VINEYARD WINERY
NICHOLASVILLE

"A Taste of History."
– Tom Beall –

There is a lot of history to discover within a few minutes' drive of this winery. Historic Camp Nelson, the Union Army supply depot which supplied over 10,000 African-American soldiers for the Civil War, Henry Clay's estate, and Mary Todd Lincoln's house are all nearby. However, the stories contained in this winery alone represent the beginnings of all commercial wineries in the United States.

The first commercial vineyard and winery in the United States was established by an act of the Kentucky General Assembly on November 21, 1799. The vinedresser for the vineyard was John James Dufour formerly of Vevey, Switzerland. The vineyard was located overlooking the Kentucky River in Jessamine County, Kentucky and was named First Vineyard by Dufour on November 5, 1798. The vineyard's current address is 5800 Sugar Creek Pike, Nicholasville, Kentucky. (wikipedia.org/wiki/American_wine)

When Tom Beall purchased property in Jessamine County during the mid-1990s, he only had one thing on his mind, an A-frame house built on a quiet plot of land alongside the Kentucky River. Always a bit of a history buff, however, he became intrigued by the idea that the first commercial winery in America might sit on his land, what he found was a much bigger story.

While researching, Beall discovered documents revealing that Daniel Boone surveyed the property in 1783 and found the Kentucky Revised Statute which established the first Kentucky

First Vineyard Winery
5800 Sugar Creek Pike
Nicholasville, KY 40356
859-885-9359

Events: Yes, outdoor weddings under
  a pavilion.

Vineyard Society. He also found connections to many others in American History including Patrick Henry and John Brown, just to name two. Deciding that such a historic place belonged to everyone, Beall quickly changed his plans for the land. He presents a tour for those wishing to learn more and see the documents signed by people from Kentucky's and America's history.

Beall re-established the terraced vineyard in the style that Dufour originally set up on the land, planting the first vines in 2008. He took special care to plant the same grape that Dufour chose for the vineyard, the Cape. He's also added Riesling, Norton, Vignoles, and Diamond grape varieties. Although his wines are just getting started, he'll soon have several types to please most every palate.

Along the way to finding his winery path, Beall established horse riding trails, spots to put in canoes, and Sugar Creek Resort, which offers overnight accommodations and event space. He took extra care to set up the tasting room in a style that is reminiscent of buildings during the period of Dufour's vineyard, adding modern conveniences. This winery is definitely worth the trip. Plan to stop by and "taste the history" of First Vineyard, soon.

177

Wines

❧ **Riesling**

Semi-sweet wine with tropical, fruity notes. This is a no fail wine that pairs with appetizers, spicy foods, pork and chicken dishes, and even desserts.

❧ **America Diamond**

Sweet and light, this is a great wine to sip while listening to an outdoor concert. Add some fruit, cheese, bread, and light cold cuts, for a picnic.

❧ **Chambourcin**

This dry, oaked red pairs well with spicy salami and beef.

Recipe

Now You See Them,
Now You Don't Wings

½ cup First Vineyard's Norton Wine

3 lbs. chicken wings, cut in two pieces with small, extra piece snipped off the end

¾ cup ketchup

½ cup apple butter

6 Tbls. hot sauce

3 Tbls. chili sauce

Boil chicken pieces for about 20 minutes. Remove from water and place on rack over a shallow baking pan. Bake at 400° for 10 minutes on each side. Mix remaining ingredients in a sauce pan, bring to a boil, and turn off heat. Add chicken to sauce to coat. Return chicken to rack and bake an additional 10 minutes. Serve with celery, bleu cheese and ranch dressing.

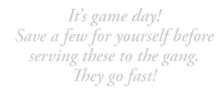

It's game day!
Save a few for yourself before
serving these to the gang.
They go fast!

CHATEAU DU VIEUX CORBEAU WINERY
DANVILLE

"De bon terroir, bon vin…" (From good soil comes good wine.)
– Dominique Brousseau –

The Chateau Du Vieux Corbeau Winery sits on the same property as The Old Crow Inn. No longer open to guests and the private residence of the Brousseau family, the Old Crow Inn is the oldest existing stone house in Kentucky. Built between 1780 and 1797, the house is on the National Register of Historic Places. Current owners, the Brousseau Family, have made restoring the property and establishing the winery a project which has lasted several years.

A cobblestone path leads past manicured gardens to the cute, tasting room. Winery owner and winemaker Dominique Brousseau comes from a long line of strong women. In the 1840s, an ancestor, Virginie de La Lallande, fought against the stereotypes of the day to become a winemaker in France. Brousseau takes her own winemaker title with a grain of salt, calling herself the "mad scientist downstairs." She's referring to her winemaking cellar. The determination passed down from her ancestors can be seen in the way Brousseau approaches her craft. "If I make something and don't like it, I stop making it. I try to make something new every year."

Brousseau is equally determined when deciding where she gets the fruit for her wines. "I use everything from Kentucky, except the cranberries. I can't convince anyone to grow cranberries." She is also very sure about the wines she makes and

Chateau Du Vieux Corbeau Winery
471 Stanford Avenue
Danville, KY 40422
859-236-1808

Events: Yes, indoor seating for up to 60 (80 if standing), outdoor, contact the winery.

when to release them. "The Cab is aged 5 years before releasing. Having patience is a must in this industry, and a lot of faith."

Dominique's parents own the Inn and her mother, Linda Brousseau, is very hands on at the winery and around the 27 acre farm. They call the barn the "PMS barn" because all of the animals, except the rooster who came free with the hens, are females. The sheep, Amazing, her daughters Grace, Emma, Lilith, and Beatrice-the diva because she tends to be aloof, are a delight to the school groups who come to visit. Usually sheep have a herd mentality and follow each other around. However, if one decides she wants to go graze by herself, she will. "My sheep have a lot of self-esteem because they do what they want to do," Linda said. Another farm animal, a goose named Miss Trouble, is not as cooperative as the sheep.

Her namesake wine's description reads; "This blend of white wines is named for the winemaker's pet goose. Much like Miss Trouble, this wine is sweet and sassy and likes to sneak up and bite you on the bottom." The lovely ladies of Chateau du Vieux Corbeau Winery, French for "House of the Old Crow,' were delightful hosts. We cannot wait to go back and taste the "mad scientist's" next creation.

181

Wines

❧ Chateau Rouge

This semi-sweet concord wine is best served slightly chilled. Spicy Indian or Latin foods will enhance its flavor.

❧ Rio Samba

This versatile wine, given a People's Choice Award at the Derby Wine Fest, is perfect to serve at everything from an outdoor picnic to an elegant dinner.

❧ Cranberry

Sweet and tart, this holiday classic is wonderful with that festive ham or turkey dinner, or adds something extra special to the holiday Wassail punch.

❧ Chardonnay

Available in both a dry and semi-sweet version, this wine is a classic treat with fruity notes and a smooth finish. Serve with white pasta sauces.

Recipe

Mike's Magic Barbeque Sauce!

¼ cup Chateau Rouge Wine
4 Tbls. diced onion
1 Tbls. olive oil
¼ cup brown sugar
Pinch of cayenne pepper
1 Tbls. honey
2 Tbls. Worcestershire sauce
½ tsp. garlic powder
1 ½ cup ketchup

In a small sauce pan, sauté onion in olive oil until soft and starting to brown. While the onion is cooking, mix the remaining ingredients in a bowl. Add mixture to the onion and simmer for 20 to 30 minutes. Remove from heat and brush on your favorite grilled meat or vegetables.

Mike's sauce is magic because any meats or veggies grilled with it instantly disappear!

PADUCAH AREA

OWENSBORO · HENDERSON · MORGANFIELD · PADUCAH · PRINCETON · HOPKINSVILLE

The Wonderland Tour

In this part of the state, there are many attractions that highlight the wonder that is Kentucky. With Kentucky Lake and Lake Barkley, plus the Ohio, Mississippi, and Tennessee Rivers, water recreation holds reign in the spring, summer, and fall. Both boat rentals and fishing guides are plentiful, even scuba diving instruction and opportunities are available. The Historic Trail of Tears played a large role in Kentucky and the Dark Patch Tobacco War, which saw the rise of the famous Night Riders, originated in southwestern Kentucky.

Both inside and outside attractions abound. Nature lovers will find much to keep them occupied here, including horseback trails, ferry boat excursions, and several water parks and golf courses. Two toy museums, several historic homes, a circus museum, and a National Quilt Museum, await visitors who come here from each state and several countries, every year.

In addition, the Western area of Kentucky offers many chances to try the local cuisine. Moonlight Barbecue in Owensboro has been voted best barbecue in the state by several magazines, Patti's 1880's Settlement in Grand Rivers received accolades from the U.S. Chamber of Commerce and *Southern Living* Magazine, and Calvert Drive-in, which resides in Calvert City, offers family movies and homemade burgers and pizza. Whatever your family might be craving, outdoor fun, indoor activities, or delicious food, there's something for everyone in this Western Wonderland.

MISTY MEADOW WINERY • *Owensboro*

RUBY MOON VINEYARD & WINERY • *Henderson*

WHITE BUCK VINEYARDS & WINERY • *Morganfield*

GLISSON VINEYARDS & WINERY • *Paducah*

PURPLE TOAD WINERY • *Paducah*

EDDY GROVE WINERY • *Princeton*

BRAVARD VINEYARDS & WINERY • *Hopkinsville*

MISTY MEADOW WINERY
OWENSBORO

Driving to the first stop on this tour, visitors will come to understand that Don Keller has been a fixture in this community for a long time. As they turn on Keller Road, they get a hint of this. Upon seeing the Keller Furniture Store, now in its thirtieth year, the Keller's longevity will be firmly established. The winery is a newer venture that opened in 2010.

"My wife (Janet) gave me a winemaking kit as a Christmas gift in 2006," Don Keller explained. She may have been influenced by his stories of a time when his grandfather and uncles, as part of their German heritage, grew vineyards on the property. The German name "Keller" means "Cellar" in English. It only makes sense that the Keller's would have a wine cellar. "I let the winemaking kit sit in the closet for a few years, then got it out to try beer making. That turned out alright, so I tried wine." What prompted him to make the leap from home wine maker to winery owner, however, was public opinion.

"I took a winemaking class at the University of Kentucky and brought in some of my homemade wine for the instructor and some classmates to taste. One guy offered me $30.00 a bottle for all of the strawberry I wanted to sell him. That's when I knew I was on to something." Misty Meadow offers 16 varieties with both grape and berry wines.

When Mr. Keller first applied for his winery license, he only made fruit wines, also called berry wines. "I think I made every mistake a winery owner can make," laughed Don Keller. "I didn't make any grape wines until someone told me I should. Then, I made red grape wines, which take longer to ferment,

Misty Meadow Winery
2743 Keller Road
Owensboro, KY 42301
270-683-0361

Events: No

so it's a longer span from production to product and reduces cash flow." Mr. Keller gained a master gardener status in 1987 and grows most of the fruit for his wines. He even has bee boxes with plans to use honey in winemaking sometime in the future.

He may be new to the winery business, but Mr. Keller knows what he's doing. When we visited, he had just entered his first winemaking competition, the West Kentucky Winegrowers Wine Competition. "I drove them over and dropped them off. They asked if I was coming back for the awarding of the medals. I told them I wouldn't come back unless I had a reason to." The Keller's have a business and farm with most of the work done by them alone. They are very busy people. "I didn't hardly get back home when they called and said it would be worth it to come back for the awards ceremony." Mr. Keller took six wines to the competition, five of them won medals. The Blueberry sweet and dry both took gold, as well as the Norton. The Chardonel gained a silver, and the mixed berry won a bronze. Very impressive for such a new winery. It's well worth it to plan a visit to this first stop on the Wonderland Tour. You may want to bring a truck to take home a new piece of furniture from the Keller Family's showroom, as well as a case or two of wine.

Wines

Blueberry

Made in both dry and sweet varieties, this one is a bestseller. Add to syrup for a special pancake or French toast treat.

Strawberry

Also made in dry and sweet, this wine was not available for tasting. It was sold out!

Mixed Berry

Made from blueberries, blackberries, strawberries, and raspberries, this wine is truly unique. Great for tasty sangria.

Norton

Dry and smooth, this wine is the perfect accompaniment to well-seasoned steaks, venison, and sausage.

Recipe

Mixed Berry Ice Cream or Anything Topping

½ cup Misty Meadow Mixed Berry Wine
1 Tbls. corn starch
1 Tbls. cool water
8 oz. frozen mixed berries, thawed and drained, reserving ½
   cup liquid
¾ cup sugar
Pinch of salt

In medium sauce pan, combine wine, reserved ½ cup liquid from thawed berries, salt and sugar. Bring to a boil, stirring to dissolve sugar. While sauce is cooking, mix cornstarch and water into a smooth paste. Once wine mixture is boiling, add cornstarch mixture. Boil 1-2 more minutes. Remove from heat, add berries. Let cool slightly to thicken. Serve over ice-cream, cakes, pies, toast or anything. This topping can be refrigerated for up to five days.

"Wine is bottled poetry."
– Robert Lewis Stevenson –

189

RUBY MOON VINEYARD & WINERY
HENDERSON

"We try to make a totally local product."
– Anita Frazer –

In 2003, Anita Frazer and Jamie Like purchased the land that would become Ruby Moon. They had forayed into some home wine making, "some good and some we had to pour out," Anita explained, and decided they wanted to open a winery. There were so many opportunities opening up in Kentucky when it came to vineyards and wineries, at that time. The mid-2000s saw a huge growth in the industry in Kentucky. Like and Frazer planted the first vines in 2004, opting for both French-American hybrids and American grape varieties. They now have three full acres of grapes including Foch, Stueben, and Niagara, just to name a few. Depending on type of grape and weather conditions, each acre can produce between two and ten tons of grapes, but that number can be higher or lower. One ton can yield about 60 cases, 720 bottles, of wine. Harvest time is a busy time of year at Ruby Moon.

Entering the lovely Tuscan inspired tasting room, guests are greeted warmly and a glass appears immediately for tasting. With some prompting, Anita Frazer tells the story of her winery path. "We visited a lot of wineries in both Kentucky and Indiana and asked a lot of questions. Dad was a tobacco farmer and I have an agriculture degree from Murray State University." Frazer became excited about the prospect of owning and working the land and developing a first class winery. "Jamie is very organized and took care of the business end of things. We

Ruby Moon Vineyard & Winery
9566 U.S. 41 A
Henderson, KY 42420
270-830-7660

Events: Yes, indoor seating for up to
60, outdoor seating for up to 100.

bottle about 10,000 bottles a year. Any successful winery needs people who are good at both winemaking and paperwork," Frazer said. They offer 14 varieties of wine, taking care to make sure each is an enjoyable experience for visitors.

"Local regulars have just become friends," Frazer laughed. It's her favorite part of owning a winery, getting to meet all the interesting people and find out how they ended up there from their corners of the world. "We've had people here from every state and 10-12 countries."

This winery really aims to please with several award winning wines, overnight accommodations, and several regular and yearly events to intrigue visitors and help the community. "Starting at 6 every Friday, we have "Fabulous Fridays" where we make up a pitcher of sangria and have cheese and crackers. There's ten percent off everything in the gift shop." It sounds like a great way to kick off the weekend. Ruby Moon also has a scavenger hunt each year for a small entry fee. The money raised goes to help Pay it Forward, a local no-kill animal shelter. In addition, they host cooking demonstrations throughout the year that teach a variety of cooking techniques, such as how to make your own pasta from scratch. Even if you're not interested in learning to make pasta, the price of admission is well worth the meal and wine pairings served. Whatever your pleasure, you're sure to find it at Ruby Moon Vineyard & Winery.

Wines

❧ **Stellar White**

This dry un-oaked wine pairs well with fish and shrimp dishes.

❧ **Moonlight Blush**

A semi-sweet blend with estate Chambourcin grapes, this wine is great for sipping or serve with ham or turkey.

❧ **Steuben**

Sweet and fruity, this wine is estate grown. Chill lightly and serve with lasagna or any favorite Italian dish.

❧ **Peach**

Sweet and smooth, this wine is great in a sangria, to sip chilled or to pair with light chicken or pork dishes.

Recipe

Easy Cheesy Dip

¼ cup Ruby Moon Steuben Wine
1 jar pineapple ice cream topping
Two 8-oz. packages cream cheese
Juice from one lemon

In microwave safe bowl, heat wine in microwave for 1 minute. Stir pineapple topping in jar and add half of the jar to wine. Reserve remaining half in refrigerator for ice cream or cake topping. Allow pineapple/wine mixture and cream cheese to come to room temperature. Mix cream cheese with fork until smooth. Squeeze juice from lemon into cream cheese, mix until smooth. Add wine/pineapple mixture and whip on low speed until smooth, about 2 minutes. Refrigerate. Serve with graham crackers and vanilla wafers. Also makes a great fruit dip. Everyone's favorite!

*"Wine is inspiring
and adds greatly
to the joy of living."*
– Napoleon –

193

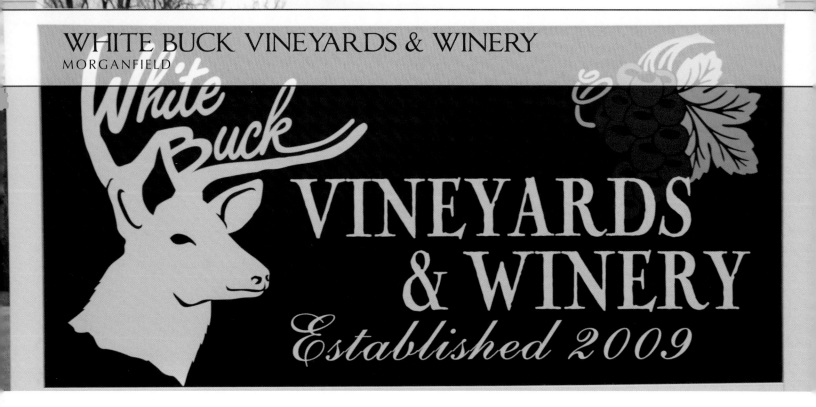

Union County is located in the Western Coal Fields of Kentucky. On their way to this gem of a winery, visitors may pass under the largest overland beltway, which transports coal to barges on the Ohio River. The land stretches out as far as the eye can see and many recreational areas are nearby including the Cave-in-Rock State Forest, Angel Mounds State Memorial, and Audubon Memorial State Park. Visitors to White Buck

Vineyard & Winery will find down home hospitality and wonderful Kentucky pride.

"We're known for wrestling in Union County," co-owner Carla White said. "We named our Sangiovese 'Wrestler's Red' in honor of this tradition." Community spirit is plain to see at this winery. "The 2008 state championship wrestling team at the high school helped us plant our first vines," Allen White announced. Employees Elaine Martin and Rob Omer showed us around the winery, made sure all of our needs were met, then spoke proudly of the owners.

"I've known Allen and Carla for quite a while," explained Martin. "My granddaughter is in school with their son." She also added, "Allen's a little modest," about his wrestling days. "Their team won several state titles while he was on it."

Carla and Allen White, both natives of Union County, have family who've farmed in the community for generations. The White's bought some land with a building they converted into a clinic. Mr. White is also known as Dr. White. They paid someone to mow the remainder of the land for a few years, then thought, "Why?" They both came from agricultural backgrounds and started investigating things to grow.

"Grapes are attractive and the wine industry had really taken off in Kentucky," noted Carla. "We planted our first plants in April of 2009, an acre and a half of French/American hybrids.

White Buck Vineyards & Winery
1800 U.S. Highway 60 W
Morganfield, KY 42437
270-389-1285

Events: Yes, indoor seating for up to
150, outdoor seating for up to 200.

They're known for their flavors and their hardiness." Since then, they've experimented with several varieties, and everything seems to be growing well. It will be a few years before the proof will show in all the grapes planted.

Even before opening, the winery has been a family affair. They named the winery after Allen's dad, "Buck" White, a fixture in the Union County community. He helped to plant the first vines and even does bottling duty, when necessary. The White's older son drew the design for the winery logo. Not to be outdone, their younger son drew a picture on a post-it note, and it hangs above the doorway in the warm and friendly tasting room. The winery offers music twice a month and these are popular events. White Buck Vineyards and Winery has a lot to offer visitors to this slice of Kentucky heaven, and they're willing to share.

195

Wines

> ### Cabernet Sauvignon
> Offered in American Oak, French Oak, and un-Oaked varieties, this dry red is fruity and bold. Serve with heavily peppered and spiced beef dishes.

> ### Sangiovese
> Described as Italy's best known grape, this semi-dry red bursts with great berry flavors. Serve with a cheesy lasagna or ravioli and marinara sauce.

> ### Dancin' Doe
> A semi-sweet white with citrus and apple notes, this wine is wonderful served chilled on a hot day or paired with lemon chicken or pork.

> ### Doc's Delight
> A semi-sweet Riesling with apple and pear tones, this wine goes well with mild cheeses and white pasta sauces.

> ### Blushing Buck
> This sweet Catawba has delicious tropical flavors. A perfect garden party wine.

Recipe

Tiramisu with Wine and Brandy

½ cup White Buck Blushing Buck Wine, divided
1 cup heavy cream
8 oz. cream cheese
5 Tbls. powdered sugar
½ cup strong coffee, at room temperature
3 Tbls. brandy
1 store-bought vanilla Bundt cake, cut into ½ inch thick slices or
    squares
½ cup semi-sweet chocolate, shaved
Cocoa for dusting

In a large bowl, combine heavy cream, cream cheese, and powdered sugar. Beat with electric mixer until mixed well. In a small bowl, combine coffee and brandy. In a 3-quart baking dish, cut cake and layer to fit bottom of dish. Slowly pour ¼ cup wine over the cake. Let the wine soak in for five minutes. (This gives you time to have a glass of wine.) Brush the soaked cake with coffee and brandy mixture. Top with half the cream cheese mixture and ¼ cup shaved chocolate. Repeat. Dust the top with chocolate powder and remaining shaved chocolate. Store in refrigerator for three hours before serving. Cover leftover and store in refrigerator for up to one week.

An easy, make-ahead recipe that goes great with a glass of sweet wine.

GLISSON VINEYARDS & WINERY
PADUCAH

A sip of Kentucky's wine history: NOW CLOSED

Paducah is hands down one of the prettiest cities in Kentucky, or anywhere. City pride is evident in the architecture, landscaping, and art that abound for visitors to experience, and much of it is free. Whitehaven, former home of Vice-President Albin Barkley, has been restored and made into a visitor center/rest stop. Murals painted on the flood wall along the Ohio River, depict scenes from Paducah's history, and one of the few museums dedicated solely to quilts, attracts visitors from around the world. In the midst of these historic and cultural areas of Paducah, sits a building that resembles an art gallery, but houses Glisson Winery.

The beautiful tile and brick exterior beckons visitors to explore the spaces within this unique structure. A tapas bar and retail shop awaits those who venture inside. The stone floor creates a path past artwork and a tile inlaid bar, leading directly to a pretty outdoor courtyard. Everything about this winery is relaxing and inviting, including the employees and owners.

Employee Ryan Mitchell was very proud of all the winery has accomplished. As he led us through the tasting, he shared his knowledge of the origins of each wine. Owners Stephen and Katie Glisson are energetic, full of life individuals. Along with owning the winery, Stephen is a father, a fireman for the city of Paducah, chief winemaker, and head grape farmer. Katie assists

Glisson Vineyards & Winery
126 Market House Square
Paducah, KY 42001
270-495-9463

Events: Yes, outdoor seating for up
to 100.

with all of the winery endeavors, plus has many outside interests, while keeping up with three energetic children.

Stephen first became interested in making wine through his friend and winery co-owner, Don Crocker. "When we first married, we bought a house that had fruit trees. We didn't know what to do with all that fruit and didn't want to see it go to waste. Katie's father suggested we try making wine." That's when their winery path began. "It took over the basement and we needed the room." Their successes led them to the idea of winery ownership.

After a few years of playing around with recipes and investigating the booming wine industry in Kentucky, Katie's father showed his belief in them by giving them some land to plant a vineyard and build a house for their growing family. After several years of marriage, the Glisson's have three children and many grape vines. "I grow the Chambourcin and Cabernet Franc grapes we use for our wine," Stephen explained. "I work with local farmers for the others needed for the winery."

When the Glisson's were ready to open the winery, Don Crocker was there too. He's part owner and "the numbers cruncher," said Stephen. "Every winery needs someone who's good at that and paperwork." Their synergy is reflected in both the building and the wines. Stop by to see and taste these perfections the next time you're in Paducah. I guarantee it's a trip you'll dream about taking again and again.

Wines

❧ **Chambourcin**

Deep red with abundant fruity flavor, this drier wine pairs well with roasted vegetables and grilled beef.

❧ **Sweet Chardonel**

Made by crossing Chardonnay and Seyval grapes, this sweet white is wonderful to sip alone or with a fruit salad.

❧ **Vintner's Blend**

One of Glisson's best sellers, this sweet red is a versatile wine to serve at any party.

❧ **Nortons**

This full bodied dry wine has hints of chocolate and thyme. Works well with a spicy beef tenderloin or grilled porkchops.

Recipe

Break-Up Cookies

¾ cup Glisson Chardonel
15.5-oz. box bakery-style sugar cookie mix or your favorite sugar cookie recipe
8-oz. jar apricot, or favorite flavor, preserves
¾ cup to 1 cup powdered sugar
3 oz. chopped nuts
½ cup shredded coconut
1 cup dipping chocolate

Make sugar cookie dough and set in refrigerator for at least 30 minutes. While it's chilling, whisk together wine and half the jar of preserves. Add powdered sugar and whisk until smooth. Add coconut and nuts, set aside. Preheat oven to 325°. Roll refrigerated dough into ½ to 1-inch balls. Using a melon baller dipped in powdered sugar or your thumb, make a small indention in each cookie. With teaspoon, fill cookies to the top with wine and preserve mixture. Bake 10-15 minutes or until golden brown. Cool completely on wire rack. Once cooled, you may glaze cookies with extra wine and preserve mixture, if preferred. Melt chocolate in microwave and drizzle over top. Sprinkle with powdered sugar, if desired. Store in single layers between waxed paper.

*A great cookie to make
when someone breaks your heart.
It will keep you busy and you can share
them with the ones you love most!*

*"Sorrow can be alleviated
by good sleep, a bath
and a good glass of wine."*
– St. Thomas Aquinas –

Oh… and a cookie!

PURPLE TOAD WINERY
PADUCAH

"If they try it, they'll buy it."
– Allen Dossey –

There's a saying that goes, "If you want to get something done, ask a busy person to do it." Purple Toad's owner Allen Dossey would certainly understand this saying as he seems to be everywhere at once. His day job is president of an insurance company. The winery only has one employee, other than him, but is open every day of the year, except Sundays and big holidays.

Many Friday evenings, you can find him in shops that sell his products conducting wine tastings and greeting customers like old friends.

"I grew up on a farm," Dossey said, "but this is gentleman farming. The vines are so pretty and the end product is so nice." As he waves to another visitor, it seems everyone who comes into the winery is Dossey's friend. If not, they soon will be. Asked about how he first became interested in winemaking, Dossey shared his story.

"I used to play a lot of golf. I took a golf trip with some friends to Napa Valley and I must have talked about it a lot. We (his wife June and himself) had an abundant blackberry crop one year and I picked them all. My wife went to the store and bought a winemaking kit for me." He entered an amateur competition and won gold on the first try. The wine bug bit hard and he's never lost a competition since.

Since opening in July of 2009, he's been rapidly expanding his vineyards and wines, now offering 19 varieties. While visiting the winery, you will often see people strolling among the vines to enjoy their beauty and symmetry.

Purple Toad Winery
4275 Old U.S. Highway 45 S
Paducah, KY 42003
270-554-0010

Events: Yes, indoor seating for up to
190, outdoor seating for up to 500.

Allen Dossey is one of those people who just cannot let grass grow under his feet. He's done quite a lot in the past to stay busy—kept bees, drawing, and stained glass art, but winemaking held his interest more than anything else. He found he liked it a lot and started to experiment with different types of wine. "Creating the perfect wine is like a successful chemistry class," Dossey explained. Finding that perfection and sharing it with others is very satisfying for him. He's a very social person, so opening a winery to the public was a natural progression of his personality, but profit isn't all he's interested in.

"I host several charity events every year for free." Easter Seals, United Way, and Cancer Society, are just a few of the charities he's hosted. "Then I donate a percentage of the receipts from the day to them. It's just a way to share all the support the community gives to me."

Once you visit Allen Dossey and Purple Toad Winery, relax in the tasting room, sip the delicious wines, and walk through the vineyards, you will want to support this winery too.

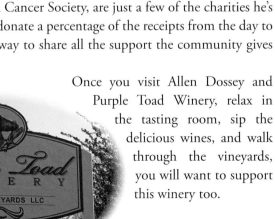

203

Wines

❧ **Pinot Noir**

This dry red, aged in French Oak, carries hints of raspberry and black cherry. Pairs well with lightly seasoned lamp and pork dishes.

❧ **Black Shadow**

A blend of Merlot and blackberry wines, this semi-dry best-seller is wonderful to sip chilled or with your favorite ham dinner.

❧ **Grant's Pomegranate**

Dark red, this wine is delightfully sweet and tart. Great wine to add to a sparkling punch or pair with Mediterranean foods.

❧ **Sangria**

A blend of sweet blackberry, peach, strawberry, and Chardonnay, this wine is a party in a bottle.

Recipe

Kentucky Chocolate Gravy and Biscuits

½ cup Purple Toad Black Shadow wine
1 ½ cups milk
4-6 warm baked biscuits
5-oz. box cook and serve chocolate pudding

Mix wine, milk and pudding mix. Bring to roiling boil. Immediately serve over warm biscuits. This wine adds a lot of depth to the chocolate flavor.

So Kentucky and so delicious!

We are all mortal until the first kiss and the second glass of wine.
– Eduardo Galeano –

Yes you can have chocolate and wine for breakfast! Serve warm with biscuits. If you should have leftover gravy, add a little or a lot of wine and mix well. Perfect for a fruit dip!

EDDY GROVE VINEYARD
PRINCETON

"Home of Medicine Man Wines."
– David Hall and Jenny Franke –

When entering the expansive grounds of Eddy Grove Vineyards, visitors feel an immediate sense of peace. The long rows of grapevines, large shade trees, one with a tire swing, and beautiful tasting room, all harken back to a simpler era when time moved at a slower pace. However, that's by design. It took a lot of work to create this image.

"I'd been a cardiologist for 25 years, explained co-owner Dr. David Hall. After multiple surgeries, my doctor told me I had to slow down. In 2001, he and his wife Dr. Jenny Franke, left their busy practices and purchased the 70 acre farm. While looking for a hobby he found relaxing, yet satisfying, he remembered their honeymoon. "We visited some of my wife's family in Italy. The whole town seemed focused on the winery, helping harvest, working in the winery, and they'd have these large scrumptious dinners that would last for hours. They told stories, jokes, sang, it was just a huge experience every evening." Hall and Franke are creating this atmosphere by hosting large dinners in their own home on the property.

The tasting room, with its wood floors and antique wine presses, has the feel of a lake cabin. Dr. Hall's father was also a practitioner and his aged doctor bag sits open as a display on top of a wine shelf. Everything about the tasting room pays homage to people and times in the life of Hall and Franke.

Eddy Grove Vineyard
300 Martin Road
Princeton, KY 42445
270-365-9463

Events: Yes, indoor seating for up to
25, outdoor seating unlimited.

Although a less stressful lifestyle is what led them down their winery path, they seriously pursued it.

"I read an online article from University of Kentucky researcher Jerry Brown about growing grapes in Kentucky," Hall said. "I invited him out to assess our farm and give advice about the best grapes to grow." They started out trying American, French-American hybrids, and the noble European viniferous varieties of grapes, as well as several types of berries. Painful experience showed them that the French-American hybrids were the best suited for their region of the state. They grow seven varieties of grapes, blackberries, and blueberries, and supplement these with some grapes from other Kentucky vineyards and some California growers. This allows them to offer a great variety of wines. In addition to wine, Eddy Grove, Home of Medicine Man Wines, offers wine related products for purchase, as well as Kentucky cheeses, teas, and coffees. Coffee-house style get-togethers are offered Friday evenings and summer concerts and festivals are planned. Although David Hall started the winery with an eye towards slowing down, it doesn't look like he'll be doing that any time soon.

Wines

✵ Crimson Cuvee

This bold, dry is a blend of Cynthiana, Red Zinfandel, and Cabernet Franc. This is the wine to serve at a backyard barbecue.

✵ Vidal Blanc

A dry white that pops with the flavors of citrus, pear, and pineapple. This silver medal winner is best served chilled with chicken or shellfish.

✵ Red Passion

Sweet and smooth, this wonderful blend of Cyntiana, Concord, and Reliance grapes is perfect served slightly chilled. Use as a sangria base for a wonderfully fruity drink.

✵ Blue Magic

This Sweet Blueberry is luscious and full of summer's flavor. Sip alone or with a favorite dessert.

Recipe

Passionate Kisses Red Velvet Cake

Ganache

1 cup Medicine Man Red Passion wine

1 cup heavy whipping cream

1 ½ cups semi-sweet chocolate chips

Pour cream into heavy sauce pan and bring to near boil. Do not boil. Place chocolate chips in glass or stainless steel bowl. Pour heated cream over chocolate. Stir with wooden spoon until chocolate is melted. Stir in wine. Place the bowl, uncovered, in the refrigerator to chill for at least one hour. Place electric beaters in freezer to chill. Just before icing cake, take chilled frosting and whip on high two to three minutes until mixture lightens in color and triples in volume.

Cake

1 cup Medicine Man Red Passion wine

Red velvet box cake mix

20-oz. can cherry pie filling

2 eggs

1 tsp. vanilla extract

Hershey Kisses, for decoration

2 generously greased 8-inch heart-shaped pans, or use round pans and cut cakes to heart shape

Preheat oven to 350°. Combine wine, cake mix, cherry pie filling, eggs and vanilla extract. Blend with an electric mixer on low speed for 1 minute. Stop the machine and scrape sides with rubber spatula. Mixture will be thick. Increase speed to medium and beat at least 2 minutes or until mixture is well combined. Divide the batter between the prepared pans. Smooth out with a rubber spatula. Place pans side-by-side in oven. Bake 30 to 35 minutes or until a toothpick pressed in center comes out clean. Remove pans and place on wire racks to cool for 10 minutes. Run a sharp knife around the edges and move to wire rack to cool completely. If necessary, shave off a little cake so the layers stack evenly. Place bottom layer on plate. Poke several holes and drizzle one to two tablespoons of wine over cake. Spread about four table-spoons of ganache over first layer. If frosting seems a bit thick, microwave a few seconds to make it easier to spread. Set the second cake on top and repeat process. Decorate with Kisses.

Cooking with wine is tricky,
I never know how much
I need to save for the recipe.

"We are dedicated to crafting quality wine made from Kentucky grown grapes."
– Jim and Jan Bravard –

Owners Jim and Jan Bravard have had their sights set on a wine path longer than many of the winery owners in Kentucky. Jim first learned to make wine while attending college and studying in Germany. A professor taught him how to make wine and he was hooked from the start.

Jan's family had always visited wineries and appreciated all they had to offer. They were interested in winemaking as well, and passed the love on to their daughter. Jan and Jim met in college and discovered this shared passion pretty quickly.

"We bought this farm within a few weeks of marrying," said Jim Bravard, and started planting in 1981." This makes them one of the first Kentucky families to consider owning a winery in the resurgence of the wine industry. The Bravard's philosophy was that they wanted to work the ground, have the vineyard on their property, and create the wine from vine to bottle. "We planted 250 vines the first time, and didn't know what we were doing", Jim Bravard said. You might say the Bravard's were as green as their vines when it came to the wine business. "Only 3 lived," Jim said.

Life marches on as it tends to, but the Bravard's never forgot their winery dream. They had three girls, worked full-time jobs, and, after some research, decided to give the vineyard another try in 1987. This time the vines survived and they opened the winery on December 12, 1992—the winery re-

Bravard Vineyards & Winery
15000 Overton Road
Hopkinsville, KY 42240
270-269-2583

Events: No

cently celebrated its 22nd anniversary. Today, the family maintains six acres of grapes. They invite visitors to bring a picnic and stroll through the vineyards or tour their quaint wine cellar. Two winery dogs may accompany visitors around the property. Reid is named for Rex Reid of Reid's Livery Winery, someone who has helped them a lot through the years. Hoss is named for Hoss Cartwright, of Bonanza fame.

Throughout the years of planting the vineyards, doing all the work required to maintain them, making wine, and introducing their wines to visitors, the Bravard's have also worked full-time jobs. Jim at Ebonite International, which makes bowling balls, pins, and other equipment, and Jan at Museums of Historic Hopkinsville, which houses three separate museums. They are the Pennyroyal Museum, which has the general history of the area, the Woody Winfree Fire-Transportation Museum, and the Charles Jackson Circus Museum—which is the only circus museum in Kentucky.

The Bravard's host several festivals each year including a harvest celebration in October and an Open House during the Holiday Season in December. Whatever you like about Kentucky and wine, you're sure to find it at Bravard Vineyards and Winery.

Wines

❧ Foch

A dry, medium bodied red wine with vibrant color and bouquet. This wine is best at room temperature and pairs well with lamb, beef, or duck.

❧ Red September

This semi-dry red is delightful and best served at room temperature. Pair with a well-seasoned roast beef.

❧ Duke's Ridge

A semi-sweet red that is their bestseller. Great to sip alone or pair with strawberry shortcake.

❧ Renaissance

A sweet white that is light and tasty as a spring rain. Serve chilled with fruit or sip alone on a hot summer evening.

Recipe

Apricot Casserole

1 cup Bravard Renaissance Wine
2 ½ cups butter crackers, crushed and divided
½ cup butter, sliced and divided
½ cup brown sugar, divided, plus 2 Tbls.
3 tsp. cinnamon, divided
Two 15.25-oz. cans apricots, drained, reserve liquid

Grease a 2-quart casserole dish. Layer 1 cup crushed crackers. Pat with ¼ cup butter, sprinkle with ¼ cup brown sugar and 1 teaspoon cinnamon. Top with 1 can of the fruit. Repeat. Sprinkle top with ½ cup crushed crackers, butter and cinnamon. Add a light topping of brown sugar. Combine the reserved apricot liquid and wine (of course we used mostly wine). Pour over casserole, adding enough liquid to be seen, if not enough, add more wine! Bake at 350° for 10 to 15 minutes. Lower temperature to 300° and bake an additional

10 minutes.

*"In wine there is wisdom.
In beer there is strength.
In water there is bacteria."*
– Old English Proverb –

Southern Kentucky Region

BOWLING GREEN AREA

UPTON • PARK CITY • ALVATON • FRANKLIN

The Cave Country Tour

In this area of Kentucky, caves are the main source of entertainment. One is recognized worldwide and attracts over 2,000,000 visitors annually. At around 400 miles of navigable passageways, Mammoth Cave is officially noted as being the longest cave system in the world and was first established as a national park in 1941. It has even gained an international Biosphere Reserve classification, for its ecosystem conservation efforts (wikipedia.org/wiki/Mammoth_Cave_National_Park). However, Mammoth is not Kentucky's only cave of note. Lost River Cave is the only underground boat tour in the state and Diamond Caverns is often described as the most beautiful.

There are other tourist attractions in this area that will encourage imaginations to run wild. A large Tyrannosaurus Rex can be seen along the interstate close to Dinosaur World and the quirky Funtown Mountain is right next door. Visitors can get a taste of Australia at Kentucky Down Under, which also includes a short cave tour, and nearby, The Corvette Museum is the only place of its kind in the world. Dedicated to the Corvette, the museum offers interactive driving simulators and a pit crew challenge. Whether you prefer the pace to be slow, fast, or somewhere in between, there's something for everyone in Kentucky's Cave Region.

NOAH'S ARK WINERY · *Upton*
CAVE VALLEY WINERY · *Park City*
REID'S LIVERY WINERY · *Alvaton*
CROCKER FARM WINERY · *Franklin*

In this scenic portion of Kentucky, the small town of Glendale gets a lot of recognition. Whether it's the Whistle Stop Restaurant with its award winning reputation or the antique and gift shops, this burg is a must see. Arguably one of the best known small towns in Kentucky, Glendale has been featured on KET and written about in numerous books, magazines, and newspapers. A short distance past this historic city is one that seems right out of Mayberry, the city of Upton.

Driving down the main street in Upton on a Saturday morning, you might see several farmers in bib overalls discussing everything from the crops to politics. Homes are neat and

scrubbed. Whitewashed fences stand bright, as if Tom Sawyer just tricked a group into doing his chores. Also in Upton is our next stop on the tour, Noah's Ark Winery.

Terry Wooden and his brother Marty were first bit by the wine-making bug when they entered their homemade wine in the Kentucky State Fair. The first year, they took home a red ribbon. They repeated their success the second year and Terry was hooked, Marty, however, wasn't as smitten. "He was ready to try his hand at some other farming venues, but I wanted to pursue more wine making competitions, Terry said.

Noah's Ark Winery
12883 Raider Hollow Road
Upton, KY 42784
502-599-3980

Since this is in a dry county, wine
    cannot be purchased here. It's
    available at Liqor Barn Outlets and
    the Douglas Loop Farmer's Market.

Events: No

While visiting their parents' graves in Upton, the brothers spotted some land going up for auction and decided to buy it. "Our great-grandparents, grandparents, and parents all lived in Hart County. We felt we should hold on to that tradition in some way," Wooden explained. Terry's great-grandfather is an important part of a Kentucky tradition, he was a trusted moonshiner. Moonshine, according Wikipedia, got its name because the operations where often clandestine and carried out by the light of the moon. Homemade whiskey was given the name "shine" and the two were combined to describe both the drink, moonshine, and the people who made and sold it, moonshiners. After prohibition stopped the sale of alcohol in the 1920s, moonshining became very popular. Many moonshiners didn't know what they were doing so blindness and even death could occur if the shine wasn't made properly. Wooden's great-grandfather gained a reputation as a reliable moonshiner, someone who cared about his work and the people who drank it. Wooden's father would have loved to see the progress his son has made. "Dad planted Concord grape vines and I feel like he would have loved to make wine." It hasn't been easy for the Woodens though as they have no farm equipment and any improvements made to the land have all been accomplished by hand.

When asked about the name of the winery, Terry Wooden laughed. "The first thing Noah did after he left the ark was offer a sacrifice to God. The second thing he did was start a vineyard. Noah was my father's name. It seemed right to start a winery on dry ground and name it Noah's Ark because that's where the ark came to settle, on dry ground." Upton is in a dry county, so no alcohol is sold at the vineyard. Terry Wooden sells his wine at Douglas Loop Farmer's Market in Louisville, where it is very popular. It is also available at Liquor Barn. Mr. Wooden makes a dry Concord wine with only 1.5% residual sugar which is very unique for a Kentucky wine. Stop by and visit with him at the Douglass Loop Farmer's Market soon.

Wines

🍃 **Concord**

A dry wine with great flavor and aroma. It's wonderful to sip alone or with a spicy meat and cheese tray.

🍃 **Vidal Blanc**

A semi-dry wine with light and fruity flavors. It's nice to pair with light chicken or fish dishes.

🍃 **Kentucky Harvest**

A semi-sweet with peach undertones, this is a wine to pair with summer fruits or create a pitcher of white sangria.

Recipe

Family Style Mac and Cheese

1 cup Noah's Ark Vidal Blanc
16-oz. package macaroni noodles
1 cup half-and-half
½ stick butter
2 cups white American cheese, may need a bit more
1 cup barbecue chips, crushed

Cook noodles in large pot according to package directions and drain. In macaroni pot, bring wine to a rapid boil over med-high heat. Turn heat to med-low, add half-and-half, butter, and cheese, stir to melt cheese and thicken sauce. If sauce is not thick enough, add cheese, ¼ cup at a time until desired thickness. Once thickened, add macaroni back to the pot, stir well to coat and heat through. Just before serving, pour into a casserole dish and sprinkle crushed chips on top. Serve immediately.

Wine is one of the most civilized things in the world and one of the most natural things of the world that has been brought to the greatest perfection, and it offers a greater range for enjoyment and appreciation than, possibly, any other purely sensory thing.
– Ernest Hemingway –

CAVE VALLEY WINERY
PARK CITY

"Enjoy the views and vistas of Cave Valley Winery"
– Nick Noble –

The next stop on this tour, Cave Valley Winery, sits close to one of the biggest tourist attractions in the world, Mammoth Cave, which boasts more than 2 million visitors per year.

Cave Valley is just getting started, as a winery, but already has a national reputation as a premier sport shooting venue. Rockcastle Shooting Center on the grounds is a world class shooting resort with 57 events held in 2013, 3 of them international competitions. They have been featured 15 times on national television and there is always something happening on the property at this winery. "We have a friendly shooting competition every month and built a cowboy action shooting range with a blacksmith shop, saloon, and other buildings you'd expect to see in an Old West town. People dress in costume and just make a day of it," co-owner Nick Noble explained. "We have 2,000 acres, a 100 guest room hotel, a shooting range, an archery range, a restaurant, a golf course, and now a winery."

On the day we visited, a race with a 5 or 10k option was just finishing. The director of customer service, Amy "Peaches" Wells, and the runners invited us into the group as if we were one of the family. This highlights the camaraderie promoted by the Noble brothers. Nick and his brother Nathan bought the property known as Mammoth Cave Resort with an eye towards expansion. "We planted our grapes in a circle pattern with an open space in the center and vines coming out in all directions.

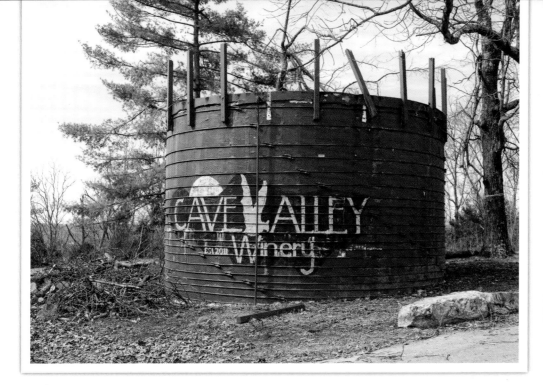

Cave Valley Winery
22850 Louisville Road
Park City, KY 42160
270-749-4101

Events: Yes, indoor seating for up to
125, outdoor seating for 800+.

It's beautiful for weddings to be held in the center." Guests sit in a circle around the bride and groom and the bride can enter through one of the many grape vine paths. "There's seating for as many guests as they would like to have. We've fed up to 800 people and we're planning an event with 1,500."

At this time, because they just received their status as a home farm winery, Cave Valley offers wines from other Kentucky wineries such as Purple Toad, Lake Cumberland, and Wight-Meyer. Jim Wight also just happens to be the vintner for Cave Valley. They are planning several wines in the future including Blackberry, Peach, Merlot, and Riesling.

Another interesting fact about the property is it sits smack in the middle of cave country. "There are 8 caves on the property," Nick said. During the winter, about 400,000 gray bats, an endangered species, hibernate in the caves. "Our future plans are to make a wine cellar out of one or more caves, as needed." Nick Noble laughed heartily at our suggestion that they nickname it "The Vat Cave." "We just might do that," he exclaimped.

Another activity on the grounds is the Cave Valley Golf Club for those wishing to work on their game. Visitors can travel the lovely countryside on one of the golf carts, pick up tips or a new club at the pro shop, or relax in the restaurant while celebrating a victory. The club also offers leagues, tournaments, and scrambles.

Come spend a weekend or two with the Nobles. Between the shooting choices, archery range, winery, golf, caves in the region, and summer concert series, plus special events, visitors won't have to leave the property to find something fun and interesting to do. Bring some friends and unwind while wine*ing* your way across Kentucky.

221

Recipe

Lemon Pork Chops in Red Wine Sauce

1 cup Cave Valley or semi-dry red wine
6 Tbls. olive oil
2 cloves garlic, crushed
6 thin-sliced pork chops
16-oz. pkg. thin spaghetti
Two 14.5-oz. can Italian-style diced tomatoes, undrained
Two 14-oz. cans beef broth
4 Tbls. flour for thickening
2 lemons

In deep skillet, add oil and garlic; turn to med-high heat. Cook pork chops until done. Remove from pan and keep warm. Add wine to the skillet and deglaze, scraping all bits from the bottom of the pan. In large pot, cook spaghetti according to package directions. Add tomatoes to the skillet and smash with fork to make a thick sauce, this takes about 15 to 20 minutes. Add beef broth and flour, whisking until smooth. Squeeze in juice of 1 lemon. Cook 10 to 15 minutes or until thickened. Place spaghetti on large platter, cover with sauce, reserving a small amount to pour over chops. Place chops over spaghetti and sauce. Pour remaining sauce over chops. Cut remaining lemon into thin slices, place over chops and serve immediately.

REID'S LIVERY WINERY
ALVATON

"Equine and Wine by Design."
– Diane Reid –

Driving south on I-65, Kentuckians realize they are in the farmer's heartland. There are a few attractions along the way, but the land quickly yields acres of corn, soybeans, and other crops essential to an agricultural way of life. Wine*ing* your way to Reid's Livery Winery, visitors have many opportunities for farmer's markets and church picnics, but their destination leads them to a playground of family fun provided by Diane and Rex Reid.

Rex's grandfather was a blacksmith and taught Rex the craft he practiced for 30 years. Rex and Diane also started a U-Pick Berry farm, but found to make any profit they "had to weigh the grandkids as they came in and left," Rex laughed. The Reids discovered that having both horse activities and the winery made it possible for them to sustain themselves on the land.

They can boast about the fact that the grandson of the great racehorse Secretariat lived on their farm. They also produce all of their own fruit, except grapes which are purchased locally, and they're one of the few Kentucky wineries producing Elderberry Wine, but there is so much more to their story.

The winery first opened in 2009 after voters approved the sale of wine on the property. The Reids worried that the vote wouldn't pass after all of the hard work they'd put into the land to produce the fruit and make the wine, and they would have to sell the wine somewhere else. Luckily, the vote easily passed and their dream became a reality.

If visitors are interested in the Livery portion of the farm, Rex and Diane offer riding lessons, boarding, training, equine dental services, birthday parties, horses for sale and even a lease program for people interested in seeing what owning a horse would be like. There are trail rides, dressage events, and a carriage for those who like getting close to horses, but not necessarily riding them.

Reid's Livery Winery
430 Nealy Road
Alvaton, KY 42122
270-843-6330

In the winery, Rex and Diane strive to make some of the best tasting wines ever made. Their Peachy Keen won a gold medal at a California competition and their Norton's won double gold in Indianapolis. Each bottle is dipped in wax, just for that little something extra that makes drinking wine a special event. Diane goes out of her way to show how to make certain foods "pop" with wine, and vice-versa. There's even a wine and ride option for those wishing to taste wine and ride trails. Whatever visitors may hope for when visiting a Kentucky winery, this one can deliver. Ride on over sometime soon.

Wines

☙ Healthy Harvest

A blend of Elderberry and Cabernet Franc, this gold medal winner is wonderful to pair with cheeses or add to a favorite red pasta sauce.

☙ Norton

This delicious dry won double gold at the Indianapolis State Fair. Serve with beef, goes especially well with grilled meats.

☙ Chardonel

A scrumptious dry white wine, this is the wine most people will buy after they comment they do not care for white wines. Works well paired with shrimp or chicken and past in an herbed cream sauce.

☙ Strawberry Patch

Made from luscious ripe berries, sipping this is like eating strawberries straight out of the garden. Mix in some strawberries and drizzle over your favorite ice cream or pound cake.

Recipe

Black Raspberry Muffins

1 cup Reid's Black Raspberry
2 ½ cups self-rising flour, plus two tablespoons for dredging
  berries
1 cup sugar
1 cup milk
2 eggs, well beaten
½ cup butter, melted
1 tsp. vanilla
¾ cup each of blackberries and raspberries

Preheat oven to 350°. Line 12 muffin tins with paper cups. Combine flour and sugar. Add milk, eggs, butter, vanilla and wine. Stir until ingredients are combined. Lightly fold in berries dredged in flour. Bake for 25 minutes or until golden brown on top. (You may want to check in 20 minutes depending on your oven). Cool in pan.

*Oh look, I got wine
in my muffins,
or did I get muffins
in my wine?
Either way, it's good!*

Drizzle

½ cup Reid's Black Raspberry
2 ½ cups powdered sugar
1 Tbls. unsalted melted butter, slightly cooled

Whisk until mixed yet thick enough to make a nice drizzle. If too thin, add powdered sugar a teaspoon at a time. If it's too thick, add wine a teaspoon at a time. Poke small holes in the muffins with toothpicks, drizzle icing over. Decorate with berries, grapes, or other fruit. Enjoy!

CROCKER FARM WINERY
FRANKLIN

Driving to this fun winery on a late winter day, we had no idea we'd be making history. We passed many cornfields and farms in this area close to the Tennessee line, to find the gravel drive that led us to Crocker Farm Winery. Upon arrival, owner and resident vintner, Dan Crocker, greeted us warmly. The Monday before our visit, the winery had been cleared to sell alcohol in Simpson County for the first time since Prohibition was enacted in 1920. We bought the first two bottles sold (legally) in the county in 93 years. The 18th amendment to institute Prohibition in the U.S. was ratified in January of 1919, but did not go into effect until 1920. After it was repealed in 1933, counties were granted the right to decide by vote if they wished to remain dry, and many did.

To the eye, the two acre vineyard may seem small, but it is mighty. Most of the vines are about 14 years old and in full production mode. One acre of grapes can produce up to about 7,200 bottles of wine, in an especially good year. The Crockers produce more than they can make and sell grapes to other vineyards and home wine makers. Full-time veterinarians Dan and Roni Crocker's winery path was set while traveling. "I was one of those people suckered into opening a winery by touring Europe as a college student," Dan joked.

The beauty of the vineyards enticed him and it didn't seem like much was happening, so he thought it would be easy to have a career and run a small home farm winery. When he has time to think about that now, he laughs and laughs. Owning a winery is hard, year-round work. Mr. Crocker doesn't sound too worried about that though, as he has plans to open a full-food service restaurant at the winery. Full-food service means Mr. Crocker will provide all of the food himself or get it locally. They already raise beef and lambs, so opening a restaurant to promote eating food from within a certain amount of mileage to your home, just seemed like a natural progression.

Crocker Farm Winery
5892 Scottsville Road
Franklin, KY 42134
270-776-2030

Events: Yes, outdoor seating for up
to 100.

Touring the farm, visitors see Dan's adeptness at repurposing materials. Timbers in the tasting room were once telephone poles, but he calls it "just being cheap. I'll take anything that's free." In addition to the cattle and lambs, there's also a trout stream on the property, barns, cows, and wide open spaces. If you're in Southern Kentucky, be sure to stop in for a sip and a laugh with the Crockers.

Wines

⚘ **Reliance**

A crisp, dry white comparison to a Riesling or Chardonnay. Works well with almost any food.

⚘ **Concord**

Sweet perfection, this wine is great to sip or adds interest when added to cakes, pies, or chocolate and strawberry desserts.

⚘ **Geneva Red**

This blend is Cabernet Sauvignon, Cabernet Franc, and Zweigelt to add body and uniqueness. Pair with a breaded, lightly fried veal and egg noodles.

Recipe

Wok This Way Shrimp Curry Stir Fry

1 cup Crocker Farm American Reliance Wine
1 cup teriyaki sauce
1 tsp. curry powder
1 tsp. turmeric powder
1 Tbls. soy sauce
2 lbs. raw, shelled, and deveined jumbo shrimp
24 oz. oriental vegetables Rice
2 Tbls. cooking oil

Mix wine, teriyaki, curry powder, turmeric powder and soy sauce in a large bowl. Marinate shrimp in wine sauce for 3 hours. In large skillet, cook vegetables according to package directions and pour into a ceramic bowl. Pour shrimp/marinade mixture into the vegetable skillet and stir for 3 to 5 minutes or until the shrimp are completely pink. Drain excess marinade; stir in the veggies and serve with rice.

Quickly, bring me a beaker of wine,
so that I may wet my mind
and say something clever.
– Aristophanes –

South Central Kentucky Region

SOMERSET AREA

EUBANK • SOMERSET • MONTICELLO • BURKESVILLE

The Lake Cumberland Tour

Wherever people drive in this region of the state, they're not far from water. Because of the humidity, fog often sits in the hollows and valleys of this part of Kentucky, creating a beautiful sight, and an eerie feeling.

Water is the main form of entertainment. Although there's an abundance of rivers and lakes, there's also a water park called Somersplash with 20 acres of slides, wave pools, and family fun. Parks and recreation areas abound including The Daniel Boone National Forest, Pulaski County Park, Rock Hollow Recreation Area, and General Burnside Island State Park—the only island state park in Kentucky. Whether you choose to be on land or water, there is a lot to keep you busy in this beautiful part of the Commonwealth.

CAVE HILL VINEYARDS • *Eubank*
SINKING VALLEY WINERY • *Somerset*
CEDAR CREEK WINERY • *Somerset*
LAKE CUMBERLAND WINERY • *Monticello*
UP THE CREEK WINERY • *Burkesville*

We arrived at our destination after dark, only to find the owners of this winery and Vineyard were ready with food, wine, and lots of jokes. Bill and Debbie Patterson were so hospitable that we got caught up in the conversation and stayed much later than we intended, but no one seemed to mind at all.

"You can't have a bad time at a winery," said Bill, as someone cracked another joke.

One of our party, Mike Woodhouse, chimed in, "It's like that Steve Martin saying, 'You can't play a sad song on a banjo.'" Though there were only 6 of us, it was a party and that's the atmosphere that Debbie and Bill like to promote at Cave Hill Vineyard and Winery.

The Patterson's started making wine, they said, "to support our habit. We like to drink wine." When they became serious

about the prospect of owning a winery, they went to the annual wine and grape conference and bought Tom Cottrell's book "From Vine to Wine."

"We've just really been impressed with how they've made themselves available," Bill said, referring to Mr. Cottrell and the people at the University of Kentucky. With over 70 wineries in the state, you would think he would be too busy to answer his phone or offer so much help, but Mr. Cottrell is driven to see the Kentucky wine industry flourish.

In the early 2000s, the Patterson's acquired the property and began the process to gain the small home farm winery status that would allow them to sell wine in dry Pulaski County. They planted their first vines in 2005 and now harvest over 1,400 plants including Cabernet Sauvignon, Chardonel, Cayuga

Cave Hill Vineyards
2115 Smith Ridge Road
Eubank, KY 42567
606-423-3453

Events: Yes, indoor seating for up to
300, outdoor seating for up to 300.

White, and Norton. Bill also shared a bit of history that we did not know about. "It's said that most of the grape vines in France came from Kentucky."

Although I could not make an accurate Kentucky linkage, according to Wikipedia, blight hit the French vineyards starting around 1863. It took many years for them to find the cause, an aphid. By then, more than 40% of the vineyards were devastated and all were sure to be affected if left unchecked. Although opposed to the idea at first, the French found the only way to fight the aphid was to graft American vines, which were resistant to the pests, onto their plants. Since Kentucky was the third largest grape producer at the time, it does make sense that many of the vineyards in France sport vines from the Bluegrass. (wikipedia.org/wiki/Great_French_Wine_Blight)

The name of the winery Cave Hill, comes from the fact that it sits on a hill the locals have dubbed "Cave Hill" for many years because it sits on a cave. There are other caves on the property, including Dumpling Cave, but they are not able to be entered. The Patterson's look forward to serving you some of their wines, and home grown hospitality, very soon. You'll leave there with a warm feeling and it won't all be caused by the wine.

Wines

❧ **Cabernet Sauvignon**

Originating in France, this dry red is bold and made to be served with food. It pairs well with a seasoned leg of lamb or pepper crusted ahi tuna.

❧ **Norton's Cynthiana**

First cultivated in the U.S. Dry and smooth, this wine is a good accompaniment to game such as deer and turkey.

❧ **Cayuga White**

A smooth and fruity wine with a crisp finish, this semi-sweet pairs well with spicy meat and vegetable dishes.

❧ **Traminette**

This semi-dry derivative of the Gewürztraminer has a slight floral finish. It is a great sipping wine or pair with foods in sweet and sour sauce or Mexican foods.

Recipe

Creamy Chicken Lasagna

½ cup Cave Hill Cayuga White Wine

2 cups diced chicken

1 Tbls. Italian seasoning

2 cups shredded mozzarella cheese (separated)

Two 11-oz. cans cream of mushroom soup

1 ½ cups milk

⅛ tsp. ground nutmeg

⅛ tsp. ground cayenne pepper

10-oz. package frozen chopped spinach, thawed
   and well drained

1 egg

15-oz. container ricotta cheese

12 lasagna noodles, cooked and drained

½ cup grated Parmesan cheese

Place chicken, wine, and Italian seasoning in crock pot and cook for 3 to 4 hours. Drain when cooked all the way through. Measure ⅔ cup of mozzarella cheese to reserve for top layer. In a medium bowl, combine soup, milk, nutmeg, and cayenne pepper; set aside. In a separate medium bowl, combine spinach, egg, and ricotta; mix well. In the bottom of a 9x13-inch baking dish, spread ½ cup of the soup mixture. Arrange 4 lasagna noodles on the mixture. Top with ⅓ of the remaining soup mixture, ½ of spinach mixture, ½ of the remaining mozzarella cheese and ½ of the chicken. Repeat the layers ending with remaining noodles, soup mixture, reserved ⅔ cup mozzarella cheese and Parmesan cheese. Bake at 350° for 40 minutes or until hot and bubbling. Let stand for 15 minutes before serving.

SINKING VALLEY WINERY
SOMERSET

Owner and vintner Zane Burton's first experience with wine was more of an experiment than anything. "I didn't drink it, but I had to sneak and make it. My mama was a teetotaler and she wouldn't have liked it." When he was 12, Burton made his first wine out of apples and elderberries, just to see if it would work. He thought it did, but no one ever drank it, so he never knew for certain how good it turned out.

Mr. Burton was a farmer by trade before becoming a winery owner. "I raised beef and tobacco. When tobacco went, I started looking to diversify. We planted the first grapes in 2000." Not sure of what would grow in this southern area of Kentucky, the Burtons planted several different varieties. "We planted Foch, Chambourcin, Norton, Vidal Blanc, Cayuga White, Concord and Niagara. The Vidal Blanc and Foch didn't work out well, so we used the small yields for blends." In addition to the grape wines, Sinking Valley Winery also carries Apple, Blueberry, and Blackberry Wines.

Many of their wine labels get noticed, but none as much as the Prohibition Repeal Red. This sweet red wine has several different pictures depicting historic events surrounding the Prohibition era including men pouring out barrels of wine in the street and Carrie Nation with her famous ax. The funniest thing many people may not realize about the time of Prohibition in the United States, 1919-1933, is that it was illegal to

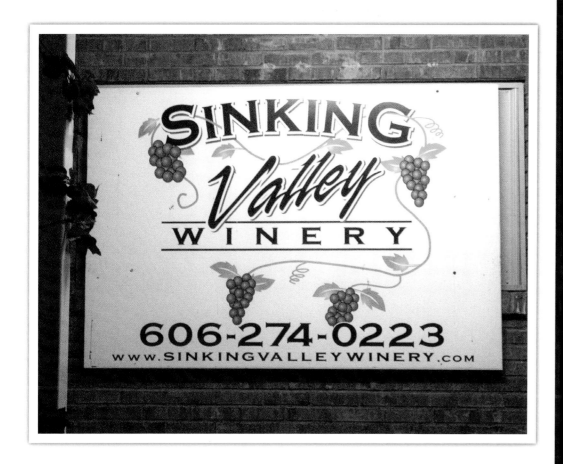

Sinking Valley Winery & Tasting Room
6515 Highway 461
Somerset, KY 42503
606-274-0223

Events: No

Sinking Valley Tasting Room
3610 Highway 27 Light #21
Somerset, KY 42503
606-451-3093

Events: No

sell wine and other types of alcohol, but it wasn't illegal to make and drink your own.

Although starting a new venture is a bit scary at first, when the idea of opening a winery occurred to him, Zane Burton plunged ahead, going door to door to get the signatures needed to put the sale of alcohol at his farm on the ballot. Once on the ballot, it passed overwhelmingly. Asked what he likes about owning a winery, he replied, "It's a pretty good business. I get to stay on the farm. I also got to buy the farm down the road, which used to belong to my great-grandfather. We got to stay in the community we grew up in. We didn't have to move to town and get a job. We get to stay out in the country." We're happy he got to stay in the country and make wines for the public's enjoyment too. We hope you'll venture out into the countryside soon and taste this little piece of Kentucky heaven.

Wines

❧ Cabernet Sauvignon

Dry and luscious with a nice density, this fruity wine carries the scents of black cherry and just a slight whiff of tobacco. Serve with rare roast beef or steak.

❧ Cumberland Blush

This semi-sweet blend is a fruit-filled burst in the mouth, producing a smooth, pleasant feel. Serve with a fruit and cheese tray.

❧ Semi-Sweet Concord

This wine from the native Concord grape is summertime in a glass. Less sweet than most, this wine is for those who prefer less sugary tastes.

❧ Sweet Apple

Fruity and crisp, like biting into a ripe Golden Delicious Apple. Pair this with your favorite fruit pie, serve with a wheel of medium cheddar cheese or sip alone. So versatile!

Recipe

How Much Cabbage for Those Peppers?

2 ½ cups Sinking Valley Prohibition Repeal Red Wine, divided

2 cups instant brown rice

1 medium white onion, chopped

½ red bell pepper, chopped

2 cloves garlic, grated

3 Tbls. olive oil

2 lb. lean ground beef or ground deer

½ tsp. seasoned salt

¾ tsp. fresh ground pepper

1 head cabbage

4 bell peppers, green, yellow, red, and orange, seeded and halved

Two 14.5-oz. cans diced tomatoes, drained

12-oz. can tomato paste

Mix ½ cup of wine with rice and allow to soak for a few minutes. In large skillet on medium high heat, cook onion, pepper, and garlic in oil for 5 to 10 minutes or until browned. Add to rice and mix in ground meat, seasoned salt and pepper; mix well. Refrigerate overnight. In the morning, remove core from cabbage and place in large pot of boiling water to remove 10 to 12 whole leaves. Hold cooled cabbage leaf in hand, place two tablespoons of meat mixture in center, fold over sides, top and bottom, and place, seam side down, in the bottom of a large crock pot. Repeat with remaining cabbage leaves. Fill pepper halves with remaining meat mixture and place on top of cabbage rolls. Pour tomatoes on top. Mix tomato paste and remaining 2 cups of wine, pour over tomatoes, peppers, and cabbage. Turn crock pot on low and cook for 10 to 12 hours. This one gets better the longer it cooks.

This crockpot recipe is so flavorful and reminiscent of my mom's stuffed peppers. I added the wine and made a few other changes. So far, no one is complaining!

CEDAR CREEK WINERY
SOMERSET

Visitors to this winery will wind down a gravel drive with quaint, antique bicycles parked along the way to point them in the right direction. Upon arrival, the scene is breathtaking with a stone and log cabin tasting room, rolling vineyards leading to a creek, and several animals milling about the property. For one in particular, a very hairy Scottish Highlander cow, guests are invited to bring a brush.

Owner and Vintner, Jeff Wiles, is a life-long resident of Pulaski County and opened for business in 2007. His mission is to "support the benefits that responsible enjoyment of wine can add to a healthy lifestyle." Mr. Wiles hopes to make the farm as sustainable as possible. He composts the by-products from the grapes during the wine making process. He adds food scraps and manure from the horses to re-feed the grapes in a continuous cycle. Mr. Wiles also catches rainwater for watering during the driest part of the summer and he's re-introduced native grasses that have not grown here since before European settlers came to this part of America. He practices organic growing procedures in the vineyard and over most of the farm. He encourages this lifestyle at his large Earth Day celebration each year. It is a day-long celebration in April with music, classes, art activities, and sustainable living demonstrations, plus wine tastings, of course.

When Mr. Wiles began experimenting with homemade wine making about 15 years ago, he had no intention of opening a winery. Even when he began seriously growing grapes, his goal was to sell them to other wineries. However, his own wine-making experiences encouraged him to try to open a winery of his own.

Cedar Creek Winery
294 Cedar Creek Lane
Somerset, KY 42501
606-875-3296

Events: No

"I started out growing red grapes because I liked the reds and I thought they would go over well. I added other wines by reading up on what would grow best here and what other wineries suggested." He started out with four wines and has gradually increased to his present ten selections. One interesting variety is called "Once in a Blue Moon." This tart and sweet wine smells like a batch of fresh-picked blueberries. Mr. Wiles is passionate about his vineyard, but he also wants to help other vineyards succeed.

"I'm on the board of the Kentucky Grape and Wine Council," he explained. The council is a government agency run through the Department of Agriculture. The purpose of the council is "to promote and facilitate the development of a Kentucky-based grape industry in the Commonwealth of Kentucky. Mr. Wiles readily shares his knowledge with friends and strangers alike. Take this beautiful drive and visit Cedar Creek Vineyards, along the banks of Cedar Creek in Pulaski County. It will be a trip you will want to pursue again and again.

Wines

Cynthiana
This semi-dry red wine has hints of cocoa and black cherry. Pairs well with anything from the grill.

Hummingbird Blush
This semi-sweet blend of Concord, Chambourcin, and Niagara grapes is a pretty amber color. It will compliment roasted vegetables, poultry, and mashed potatoes.

Marĕchal Foch
Only limited availability on this French hybrid with bouquets of plums and black cherries. Pair with red pasta sauces or even chili.

Niagara
This wholly American wine, developed in Niagara County New York, is a sweet and fruity white. Great for sipping or pair with soft cheeses and fruit.

Recipe

Easy Hummingbird Cake

¾ cup Cedar Creek Humming Bird Blush Wine, divided
18.25 oz. yellow cake mix (Do NOT use a cake mix with
    pudding.)
20-oz. can crushed pineapple in juice, undrained
2 very ripe bananas, mashed
10 maraschino cherries, chopped, reserving liquid
Reserve a few cherries for decorating
1 tsp. vanilla extract
1 tsp. cinnamon
1 can cream cheese frosting
Powdered sugar
Pecan halves

Preheat oven to 350°. In a large bowl, mix all ingredients except powdered sugar, ¼ cup wine, frosting, pecan halves, and cherries for decoration. No need to use an electric mixer; just be sure to mix well by hand. Divide mixture equally between two greased 9-inch round pans. Bake both pans on middle oven rack for 30 minutes or until a toothpick inserted into center comes out clean. Place pans on wire rack and run a knife around edges. A tip my mom taught me to get cake out of pans is to tap the bottom of the pan with a butter knife until you feel confident it will not stick. While the cake is cooling, place half of frosting in a small bowl. Add 4 Tbls. cherry juice, ¼ cup more wine and 1-2 cups powdered sugar. Stir until thickened and refrigerate until ready to use. Place one layer on a plate and frost the top with the cherry/wine icing. Place the other layer on top and frost entire cake with the remaining cream cheese icing. Decorate with pecans, cherries, or whatever you like.

*Hummingbirds, here one second
and then they're gone.
A flash of color I watch them fly,
flower to flower then off to the sky.*

LAKE CUMBERLAND WINERY
MONTICELLO

Norrie Wake, owner of Lake Cumberland Winery doesn't have a motto like a lot of the wineries, but if one were chosen for him, it would be, "When everyone succeeds, I'm successful." Mr. Wake is one of those individuals every county needs. He's involved in all kinds of community service organizations, and his winery even reflects this spirit. Not only does he have his own wines, created by winemaker Jim Brady, Mr. Wake carries wines from wineries close to him and some that are far away. These wineries include Lovers Leap, Up the Creek, Horseshoe Bend, Sinking Valley, and Talon. Visiting this one winery is a small tour of Kentucky's wine growing regions.

Mr. Wake came to be a winery owner quite by accident. He read about others trying to grow grapes and decided he would try it. "It was an experiment that went awry," he said. "I planted 400 vines at first, then found I had to fight the deer, raccoons, and possum to get a crop." After 3 years of fighting and reinforcing fences, he produced his first crop. "There were so many beautiful grapes I didn't know what to do with them all." He called in several friends to help pick and they divided the grapes. These friends have since been called the "Fermenting Friends" and they're very involved in the winery every season. Success though, can bring unforeseen problems.

"I got overwhelmed with enthusiasm when those first 400 plants produced a crop, so I planted 600 more." This is when

Lake Cumberland Winery
122 Cedar Lane Farm
Monticello, KY 42633
606-348-5253

Events: Yes, indoor seating for up to
60, outdoor seating for up to 500.

the idea of opening a winery first tickled Norrie Wake's brain. He did some research about the steps to opening a home farm winery. One of the biggest obstacles was getting the vote to sell alcohol. Once the vote was gained, he opened his winery up to other wineries as a place to sell their wines. By gaining recognition for them, he gets a compassionate reputation, and thus gains respect and admiration for many of his endeavors in the community. It's a win-win situation.

He named the vineyards, which sit nearby and are open for tours by appointment, Cana Vineyards. This is representative of the first miracle Jesus performed in the Bible in Cana of Galilee while at a wedding ceremony, Jesus turned water into wine.

The winery, vineyards, and the Lodge, a lovely home next to the winery which can accommodate 24 overnight guests, all sit on Cedar Lane Farm.

Norrie Wake's parents first bought the property on the banks of the Cumberland River. "My dad always dreamed of owning one of every type of farm animal out there, so they bought the farm and started a menagerie," he laughed. "We should have listened to the local farmers who told us don't own anything you have to feed."

Norrie Wake invites you to come and have a glass, or bottle, of some Kentucky wine and watch the sunset. Bring some friends, a picnic basket, and a few blankets, it can get cold by the river at night. Breathe in the country air and relish a place that is just this side of paradise.

Wines

◦ **Villard Noir**

The cute label on this wine gives no hint of the serious wine within. Semi-sweet and fruity, this wine pairs with a fruit and cheese tray, meat tray, or grilled burgers.

◦ **Triad**

This dry red blend has predominantly cherry and spice flavors. it pairs well with a grilled steak or Mediterranean spiced dishes.

◦ **Cayuga White**

Straight from the vine, this crisp white is a summer's day in a glass.

◦ **Vidal Blanc**

Hints of citrus and melon aged in light oak to pair well with veal or pork.

Recipe

Wine-J-Hello

2 cups Lake Cumberland Winery Villard Noir or Cayuga
  White
⅔ cups sugar
3 Tbls. unflavored gelatin
4 Tbls. Water

This dessert can be made in any shape or size you wish. We suggest lining a 9-inch square pan with plastic wrap that covers the bottom of the pan and hangs over the edge for easy removal. Combine wine and sugar in a sauce pan and bring to a boil, stirring until sugar has dissolved. In a bowl, sprinkle gelatin into the water and let it sit for 1 minute. Slowly pour the wine mixture over the gelatin and stir until completely dissolved. Cool to room temperature. Pour gelatin mixture into pan and refrigerate for 4 hours. Pull the gelatin out of the pan using the plastic wrap hanging over the edge of the pan. Use a cookie cutter or knife to create your shapes. An elegant addition to a cheese plate while on a picnic at the lake. This wine Gelee pairs well with goat cheese, Brie and of course any of your favorites.

Good food, great friends,
and wine.
A perfect night,
as long as someone remembers
to bring a corkscrew.

"We hand craft each wine!"
– Gary Haddle –

The area between the Cascade Mountain Range and northern Washington and Oregon beaches encompasses some of the most beautiful scenery in America. It is also known as a very cold region, with the Pacific Ocean rarely warm enough for swimming, so it may be surprising to learn it contains a thriving wine region.

It was in this northwestern region of the U.S. that part owner of Up the Creek Winery, Gary Haddle, along with his brother David, first became interested in wine and wineries. "I became an enthusiast and enjoyed touring wineries in Oregon and Washington. The Cascades and northern beaches reflect

a beautiful world." This was in the early 1990s, when Haddle worked in the area. He knew then he wanted to own a winery and started looking for ground in that part of the country, however, something turned his path toward Kentucky. "My brother would only join me if we shopped closer to our old home."

"We bought this farm in a Cumberland County valley known as 'Possum Hollow' during the fall of 2002 and planted the first vines in the spring of 2003." They began the vineyards before they considered anything else, not adding a cabin until later. "We take pride in our vineyards because good wine begins there. The valley pleases one's soul."

Photo credit, Gary Haddle.

Up the Creek Winery
932 Norris Branch Road
Burkesville, KY 42717
270-777-2482

Events: No

Gary is known to wax poetically about the vineyard, and his words are philosophically deep. "It's all about wine and everybody has unique senses of sight, smell, and taste. The people I meet and work with in this business maintain positive energies. I enjoy the experience of working with good folks. And of course being a man aware of endlessness, and time, and the universe, I love the mathematics and chemistry of wine making. And I like tasting good wine."

It's not all about good wine as part of the pleasure the brothers gather at this quiet, rural spot comes from the abundance shown to them from the natural world in which they participate. Gary continued. "Being here on the farm means being witness to nature. We see events like three bald eagles diving in above the Chambourcin Vineyard while working on a cold wintery day; like three falling stars on a crisp, clear night; like three deer watching from across the foggy field during the daybreak hour."

The Haddle brothers work hard to be very hands on in every part of the wine process in the vineyard, making, and bottling of the wine. Visitors may also purchase their wine at Lake Cumberland Winery from their good friend, Norrie Wake. At Up the Creek Winery, "we work hard to ensure well-balanced, quality wines. We monitor each wine and make a decision based on what that wine tells us." Do yourselves a favor and taste what the wines at Up the Creek had to say very soon.

Photo credit, Gary Haddle.

Wines

❧ Naked American Chardonnay

This wine gets its name because it is fermented in stainless steel instead of oak. A slightly sweet finish pairs well with tropical fruit or buttered lobster rolls.

❧ Muse Blackberry Wine

Made with sweet fruit picked on the farm, this semi-sweet wine has a flavor unique to this area of the country, including a hint of maple.

❧ Bramble Jazz

Blackberry and raspberry complement each other perfectly in this blended treat.

❧ Rosita Caliente

A brilliant blend of raspberry, cranberry, and strawberry. Serve this bold wine with spicy Thai food or smoked salmon. Perfection!

Below: Photo credit, Gary Haddle.

*Enjoy the little things, for one day you may look back
and realize they were the big things.*
– Robert Brault –

Recipe

Kentucky Hot Brown-ies

Brown-ies
2 slices thick cut turkey from the deli
1 tomato diced
3 slices bacon, cooked and crumbled
6 slices large bread, cut in half

Sauce
¼ cup Up The Creek Rosita Caliente Rosé Wine
½ cup half-and-half
¼ cup shredded Cheddar cheese
½ cup shredded white American cheese

*A classic Kentucky dish created
at Louisville's Brown Hotel in the 1920s.*

Combine turkey, tomato, and crumbled bacon in a small bowl. Spray cooking spray in a 12-cup muffin pan. Arrange ½ slice of bread in each muffin cup, sealing edges together to form a cup. Turn on broiler. In a small sauce pan, heat wine to rapid boil and reduce heat. Add half-and-half and cheeses, stir to melt. Keep hot. Put muffin pan in oven to toast bread. When toasted, fill with a spoonful of turkey, bacon, tomato mixture and divide cheese sauce evenly over the hot browns. Place under broiler, 1 minute to heat through. Serve immediately.

Eastern Kentucky Region

MOREHEAD AREA

GRAYSON · MOREHEAD · SECO

The Daniel Boone Tour

Although this is called The Daniel Boone Tour, it could have just as easily been called the Hills and Hollows Tour. This part of Kentucky is the most mountainous portion of the Commonwealth and is known for coal and country music. Many country music stars came from the eastern region of the state, including, just to name a few, Loretta Lynn, Patty Loveless, Dwight Yoakum, Crystal Gayle, and Billy Ray Cyrus, yes, Miley's dad. For the final winery, we travel along historic Highway 23, The Country Music Highway, but for the first two, we stay a little west of there, down Interstate 64 through the Daniel Boone National Forest.

Daniel Boone, an early frontiersman, forged the first navigable trail across the mountains of North Carolina, through part of Tennessee and into Kentucky, which was then part of Virginia. He founded one of the first American settlements west of the Appalachian Mountains, Boonesborough, along the banks of the Kentucky River in Madison County (wikipedia.org/wiki/Daniel_Boone). His supposed grave can be seen in Frankfort Cemetery, a peaceful site high on a hill overlooking the Kentucky River. I say "supposed" because there is some controversy surrounding Boone's resting place. At the time of Boone's death in 1820, he was buried next to his wife, Rebecca, who had died in 1813. They resided in Missouri at the time. In 1845, a Kentucky delegation ordered a disinterment and reburial of the bodies of Daniel and Rebecca in Frankfort, Kentucky. No positive DNA analysis has been done to date and both states still claim the resting place of the Boones (wikipedia.org/wiki/Daniel_Boone).

In Eastern Kentucky, the great outdoors is a huge pull. Hiking, rock climbing, caving, and lake activities are the main sources of entertainment. However, some friends would agree that watching the sun go down while sitting around a campfire with good conversation and a bottle of wine, or three, could also be counted as great entertainment.

ROCK SPRINGS WINERY · *Grayson*
CCC TRAIL VINEYARD & WINERY · *Morehead*
HIGHLAND WINERY · *Seco*

ROCK SPRINGS WINERY
GRAYSON

"Quality Kentucky wine."
– John and Sue Bond –

This winery has the distinction of being the farthest east in Kentucky. Down Interstate 64 past Morehead to a little beyond Mount Olive, we encountered some hard working Kentuckians, striving to make a honeymoon dream come alive and help others see their vision.

After they married several years ago, John and Sue Bond took a honeymoon trip to Lake Tahoe and toured several vineyards and wineries in the area. It's easy to get caught up in the beauty and serenity of the area. When many are faced with the reality of the amount of work required for a successful vineyard, they find another line of work, like digging ditches or construction. Not the Bonds, they seem to thrive on the work.

"We bought the barns from some tobacco farmers," Sue said as she pointed down the hill to the event barn. "It needed some work, but we didn't want to put new boards in to replace the old ones, so we found an old barn that needed taking down and offered to take it down so we could use the wood." Repurposing projects can be seen in many places. An old corn crib has been re-built to be used as a seating area for visitors or an archway for a couple to stand under as they marry. Herbs grown on the farm are used at dinners held the 1st and 3rd Fridays of each month. The lovely tasting room can hold up to 40 people and space fills quickly.

While Sue and John have had their winery dream since they

Rock Springs Winery
1081 Rock Springs Road
Grayson, KY 41143
606-474-2315

Events: Yes, indoor seating for up to
200, outdoor seating for up to 200.

married, it took them a few years to take the plunge. After buying the land with the barn, they built a house for her father, who originally lived in Jamaica. He wanted to live on a farm and raise animals, so he started collecting his menagerie. The Bonds would come on the weekends to help with upkeep and to plant the vineyards. "Originally, we planted a field of Cabernet Sauvignon and intended to sell to other wineries. After investing so much work in it, and with our line of work, opening up the winery was a natural fit." In order to get the small farm wintery status needed to sell wine in dry Carter County, the Bonds went door to door to get enough signatures to put the vote on an election ballot. Once that passed, they were prepared to open to the public in May 2010.

After living in Kentucky for a few years and experiencing our winters, Sue Bond's father decided to move to Florida. "We really miss him. We also miss coming to the farm and having fresh eggs and meat, there's nothing like it." One thing Rock Springs does differently than a lot of wineries is they grow eating grapes for picking. "That's really special," said Sue. "Many children and adults have never had that experience, to go into the vineyards and pick their own grapes to take home. They seem to really love it." We loved it too. This beautiful winery with its quiet beauty and natural resources is a great place to pass some time on a warm afternoon.

257

Wines

❧ Syrah

Full-bodied and dry with peppery hints, this wine pairs well with barbecued pork, Asian flavors, and smoked meats and vegetables.

❧ Traminette

Derived from the German Gewürztraminer, this crisp white works well with Mexican and Chinese cuisines.

❧ Plum

This sweet dessert wine is a delight to sip alone or pair with your favorite fruit tart.

❧ Honey Mead

Often considered the ancestor of all fermented drinks, honey mead is a mixture of honey and water, sometimes heated, to produce an alcoholic content that falls between 8 and 18%. Take it easy with this one, it's potent!

Recipe

Plum Tuckered Squares

¼ cup Rock Springs Winery Plum Wine
1 cup plum jam
1 ready-made pie crust

Press pie crust into an 11x7 baking pan. Cut edges, leaving a small edge all the way around, reserve cuttings. Crisp the crust by baking for 5-10 minutes before adding topping, do not over bake. In a small sauce pan, bring wine to a rapid boil. Reduce heat and add jam, cook and stir to melt. Pour over crust. Use extra dough cut in strips to decorate the top. Bake at 350° for 10 to 15 minutes until browned. Let cool before cut-ting into squares.

These dessert bars are so easy that you won't mind making them, despite how tuckered out you might be!

CCC TRAIL VINEYARD & WINERY
MOREHEAD

A sip of Kentucky's wine history: NOW CLOSED

Within the boundaries of the Daniel Boone National Forest, sits a unique winery where unexpected treasures await. CCC Trail Vineyard and Winery, also known as Celebrations Café at CCC Trail, is filled with eclectic antiques. On the walls hang several original paintings from the owner of the winery, one of them painted when he was just 16 years old. Large mirrors reflect the mix of objects all around and a suit of armor stands guard over the tasting room. Downstairs, an unusual wine bottle window is situated in a room filled with stained glass windows. Holding court over this entire conundrum is an eccentric character of a proprietor named Jim Ross.

Although he's a winery owner now, he never intended for that to happen. "Sometimes it's just the right time to try something new," said Ross. His first wine experiences were while in college to become a doctor. "A friend talked me into going on a trip to San Francisco, then to Napa Valley." After that, he didn't really have much wine-making experience, until he bought a winery. Much of Mr. Ross's desire to own a winery involves food.

"Part of my passion is to educate people about regional foods. People seem to enjoy stories around their food." He and winery assistant manager Robin Hanshaw, have discovered this during their weekly Saturday evening dinner parties. These dinners are open to the public, but reservations are required. "I

CCC Trail Vineyard & Winery
3939 CCC Trail
Morehead, KY 40351
606-780-7195

Events: Hosts a Saturday evening dinner every week by reservation. No other events.

cook, Robin serves," Ross explained. "We source locally for as much of the food as we can. People love the idea of having a meal where everything is made here or from local farmers." The dinners are a seven course affair that last up to 3 hours. "The main complaint is there's just too much food, so we pack up leftovers for them to take home with them. It's all about making people happy and giving them an experience."

Experiences are what Ross hopes to give all patrons wishing to visit CCC Trail Vineyard and Winery. In addition to the winery, he also owns an inn nearby, Journey's Inn Lodge. He is acquainted with rock climbers and fishing guides for his guests enjoyment. In addition, for those who do not enjoy wine, but love the winery's surroundings, afternoon teas are held at the winery.

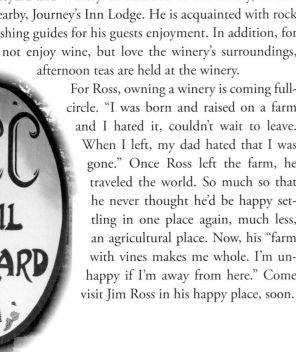

For Ross, owning a winery is coming full-circle. "I was born and raised on a farm and I hated it, couldn't wait to leave. When I left, my dad hated that I was gone." Once Ross left the farm, he traveled the world. So much so that he never thought he'd be happy settling in one place again, much less, an agricultural place. Now, his "farm with vines makes me whole. I'm unhappy if I'm away from here." Come visit Jim Ross in his happy place, soon.

261

Wines

❧ Rowan County Red

A dry French hybrid, this wine has notes of black cherry and currant. Pair with a spicy roast beef or Indian cuisine.

❧ Rowan Ridge Red

This American blend of Catawba, Niagara, and Concord is a sweet treat lightly chilled. Serve at an afternoon garden party with cucumber sandwiches.

❧ Vidal Blanc

Bursting with fruity flavor, this wine carries notes of both grapefruit and pineapple. Serve with white lasagna or other white pasta or cheese sauces.

❧ Golden Harvest

This award winning sweet wine has a stunning golden color and burst of flavor. Chill slightly and serve with your favorite dessert.

Recipe

Red Russian Roast

2 cups CCC Trail Rowan County Red
8-oz. bottle Russian dressing
3 to 4 lb. beef roast
2 cans sliced carrots, drained
2 cans whole potatoes, drained

In large, heavy bottomed pot, pour small amount, about ½ cup of Russian dressing and turn on medium high. Brown roast in dressing, turning to get all sides. Add the remaining dressing, wine and enough water to cover ¾ of the roast. Turn heat down slightly, cook for 3 hours, or until fork tender. Add carrots and potatoes, cook until heated, serve with crusty bread and more CCC wine. Note: For a crock pot recipe, brown roast and add to crock pot. Cook on low with wine, water and remainder of dressing for 8 to 10 hours. Add potatoes and carrots and turn on high until heated through.

I cherish the moist corks,
freshly picked tomatoes,
warm homemade sourdough bread
and enough sweet smell
of puppy breath to ensure
I am enjoying every minute
God grants me to be alive!
– Jim Ross –

Worth the Trip...

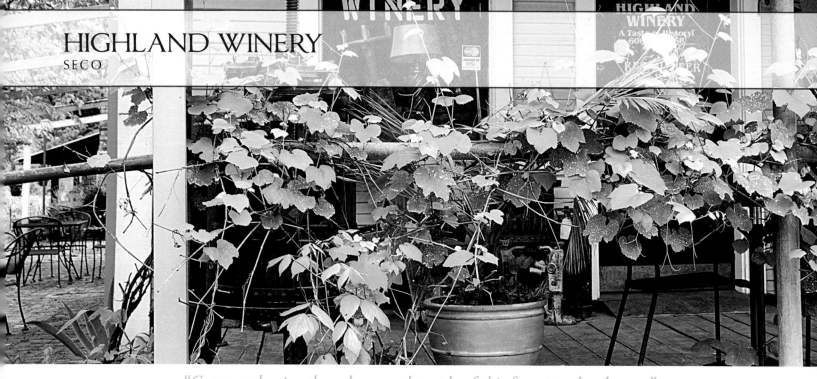

HIGHLAND WINERY
SECO

"Come and enjoy the culture and people of this forgotten landscape."

To get to this part of Kentucky, visitors travel U.S. Route 23, also known as the Country Music Highway because of the dozen or so country music stars who were born in the area. Deemed a National Scenic Byway in 2002, this portion of the commonwealth contains some of the most mountainous land in the state, which affords some picturesque vistas. Highland Winery is located in far eastern Kentucky, near the border of Virginia, and doesn't have another Kentucky winery close by. This winery may be all alone, for now, but it is definitely worth the trip. In addition, owners Jack and Sandra Looney have room at their inn for 56 guests, so bring your friends along on this adventure, just be sure to let the Looney's know you're coming.

Visitors who arrive unexpectedly at Highland Winery may find a phone on the windowsill of the expansive front porch along with a note explaining the owners are busy in the vineyards and please call one of the numbers listed. Driving up to the pretty yellow building that houses Highland Winery, visitors may feel as if they've traveled back in time. The winery used to be the company store for the South Eastern Coal Company and the Looney's have stayed faithful to its history as they reconstructed the building from memory. The name of the town, Seco, even came from the mine.

Mr. and Mrs. Looney became winery owners in a roundabout way-their daughter became very interested in the process of growing grapes and making wine. When she decided to move to Lexington and open the Jean Farris Winery, she left behind, according to Mr. Looney, "about 5,000 gallons of wine. But

Highland Winery
193 Seco Drive
Seco, KY 41849
606-855-7968

Events: Yes, indoor seating for up to
350, outdoor seating for up to 350.

that's just an estimate." Mr. and Mrs. Looney, like many thrifty Kentuckians, could not let so much go to waste, so they started their own winery. They now bottle 27 varieties of wine, but the funny thing is they do not consider the winery their main business. Mr. Looney has another full-time job and Mrs. Looney loves the farm life with her milk cows and gardens. Another thing that makes this winery definitely worth the trip is that there are three other wineries nearby, they're just all in Virginia. Visiting in late spring, snow seems to be in the air, but in actuality, it's all the blossoms from the apple, peach, pear, and cherry trees on the property as they prepare for their abundant harvest. Whichever time of the year you plan a visit, the Looney's make it worth the trip, definitely!

Wines

❧ Peach

Unusually dry, this wine has a light, not-too-sweet taste from start to finish. It pairs well with both Havarti and camembert cheeses and a variety of fruits.

❧ Merlot

"The best Merlot in the state of Kentucky. That's no brag, just fact." Mrs. Looney explained. Smooth and slightly oaked, this wine pairs well with almost any beef or lamb meal.

❧ Maiden's Blush, Mountain Maiden, and Maiden's Pleasure

Semi-dry, semi-sweet, and sweet, respectively. Each is distinctive and worth the sip.

Mr. Looney is most proud of their Miner's series.

❧ Miner's Blood

A sweet red and their bestseller.

❧ Miner's Sweat

A sweet white with a burst of pear taste.

❧ Miner's Tears

A smooth white with a light honey finish.

These wines would pair well with foods containing cinnamon and/or nutmeg, everything from chili to apple pie.

Recipe

Chocolate-Wine Balls

¼ cup **Highland Winery Miner's Blood**
6 oz. Semi-sweet chocolate chips
2 ½ cups finely-crushed vanilla or chocolate wafers
¼ cup honey
2 cups ground walnuts
1 cup sugar in a small bowl

Heat chocolate chips and honey in a 3 quart sauce pan over low heat, stirring constantly, until chocolate is melted. Remove from heat. Stir in crushed wafers, walnuts, and wine. Shape into 1-inch balls; roll in sugar. Store Chocolate-Wine Balls in a tightly covered cookie jar or metal can. Let stand for several days or even up to 4 weeks. Flavor improves with age!

This recipe is so easy and delicious it makes everyone smile!

Up and Coming Wineries

BENTGATE FARMS

336 Cook Road
Mt. Vernon, KY 40456

BOUCHERIE WINERY

6523 Keyway Drive
Spottsville, KY 42458
270-826-6192
Now Open

BRIANZA GARDENS & WINERY

14611 Salem Creek Drive
Crittenden, KY 41030
859-445-9369
Now Open
Events: Yes, call for details.

CARRIAGE HOUSE VINEYARDS

259 Longview and
Auburn, KY 42206
270-847-1519

CHRISTIANBURG FARMS

264 Christianburg Road
Shelbyville, KY 40065
502-461-9320

ECHO VALLEY WINERY LLC

1662 Mount Caramel Road
Flemingsburg, KY 41041
606-845-0030
Now Open
Events: Yes, call for details.

FANCY FARM VINEYARD & WINERY

177 Hayden Street
Fancy Farm, KY 42039
270-623-8412

GREEN PALACE MEADERY

244 Old Edmonton Road
Glasgow, KY 42141

HAMON HAVEN WINERY

7041 Rockwell Road
Winchester, KY 40391
859-745-4161
Now Open

HERITAGE AT KENTUSCANY

669 Roachville Road
Campbellsville, KY 42718
270-789-3636
Now Open, by appointment only.

LONG LICK FARMS WINERY

6769 Polin Road
Willisburg, KY 40078
859-284-1002

LOUISVILLE MEAD COMPANY

2005 Douglas Boulevard
Louisville, KY 40205
502-210-9974
http://louisvillemeadcompany.com/
Saturdays from 10-2.
Available at Louisville's Douglas Loop Farmer's Market

LULLABY RIDGE

1020 Ephesus School Road
Waynesburg, KY 40489
606-355-0039

MATTINGLY FARMS WINERY

2033 Thomerson Park Road
Austin, KY 42123
270-434-4124
Now Open

OAK LANE VINEYARDS & WINERY

3155 Keene Road
Nicholasville, KY 40356

SILVER SPRINGS FARM

3710 Leestown Road
Lexington, KY 40511
859-351-9067

About the Author

Becky Kelley has two Bachelor of Arts degrees in English and psychology from Spalding University. She also received a Master's Degree in English from the University of Louisville, and spent two years in a doctorate program of psychology at Spalding.

She's written many non-fiction pieces published in numerous newspapers and magazines such as the *Courier-Journal*, *Southwest Reporter*, and AAA's magazine *Home & Away*, as well as, *Reader's Digest*, and *Kentuckiana Family*. She has a self-published children's book, *A Tail of Christmas*, and she has a mystery fiction series in production. In addition, she's won many contests including 1st place in the Art of Crime contest, 2002, 2nd place in a collaborative university contest held between eight different colleges, 2nd place in the Bullitt County Library Writing Competition, 2014, as well as many honorable mentions in local and national contests.

About the Photographer

Kathy Woodhouse has been photographing, professionally, for more than three decades. Her experience has spanned a myriad of styles including studio, retail and managing multi-unit retail where she taught photography, managed personnel and trained studio management. Amongst her many accolades, Kathy was named National Sears Portrait Studio Manager of the Year; her work was featured on NBC's *Today Show*, December 13, 2013. For the last decade, Kathy has owned her own studio and worked as a freelance photojournalist, marketing her work at various galleries and retail outlets in Kentucky.

Kathy is a proud native of Kentucky and Bullitt County and is a graduate of North Bullitt High School. She also attended Elizabethtown Community College and is certified in both photography and Graphic Arts.

Index of Wineries

Index of Recipes

Sauces

Side Items

Soups

Sweets